DATE DUE

DEC APR 1 3 1996	APR 2 6 1995
DEC 2 1 1995	MAR 2 8 2002
DEC 1 8 1995	APR 2 8 2003
APR 2 6 1998	APR 2 8 2003
APR 2 2 1996	
JUN 1 3 1996	OCT 1 9 2005
DEC 1 8 1997	DEC - 6 2005
JAN - 9 1998	MAR - 2 2006
FEB 2 6 1998	DEC - 3 2008
MAY - 6 1998	NOV 2 4 2008
NOV 1 3 2001	
NOV 2 8 2001	
MAR 2 7 2002	
APR - 2 2003	
APR - 2 2003	

Controlling Reproduction

THE INSTITUTE OF BIOLOGY AND CHAPMAN & HALL

Biology is a disparate science, embracing a spectrum of study from the molecular level to groups of whole organisms. It is increasingly difficult for both professional biologists and students to keep abreast of developments. To help address this problem the Institute of Biology is complementing its own in-house publications with those produced by professional publishing houses. The Institute is pleased to be publishing *Controlling Reproduction* with Chapman & Hall. The Institute's Books Committee welcomes proposals for other mid-university level texts.

THE INSTITUTE OF BIOLOGY

The Institute of Biology is the professional body for UK biologists. It is a charitable organization, charged by Royal Charter, to represent UK biology and biologists. Many of its 15 000 members are Chartered Biologists (CBiol), a qualification conferred by the Institute on professional biologists and which is recognized throughout the European Communities under Directive 89/48/EEC.

The Institute's activities include providing evidence on biological matters to government, industry and other bodies; publishing books and journals; organizing symposia; producing specialized registers and coordinationg regional branches. The Institute is a prominent member of the European Communities Biologists Association (ECBA) and co-ordinates liaison between UK biologists and the International Union of Biological Sciences (IUBS).

For further details please write to:

The Institute of Biology
20–22 Queensberry Place
London
SW7 2DZ

Controlling Reproduction

J.S.M. Hutchinson

Department of Agriculture
University of Aberdeen
and
The Rowett Research Institute
Aberdeen
UK

The Institute of Biology

Incorporated by Royal Charter

CHAPMAN & HALL

London · Glasgow · New York · Tokyo · Melbourne · Madras

Published by Chapman & Hall, 2–6 Boundary Row, London SE1 8HN

Chapman & Hall, 2–6 Boundary Row, London SE1 8HN, UK

Blackie Academic & Professional, Wester Cleddens Road, Bishopbriggs, Glasgow G64 2NZ, UK

Chapman & Hall Inc., 29 West 35th Street, New York NY10001, USA

Chapman & Hall Japan, Thomson Publishing Japan, Hirakawacho Nemoto Building, 6F, 1-7-11 Hirakawa-cho, Chiyoda-ku, Tokyo 102, Japan

Chapman & Hall Australia, Thomas Nelson Australia, 102 Dodds Street, South Melbourne, Victoria 3205, Australia

Chapman & Hall India, R. Seshadri, 32 Second Main Road, CIT East, Madras 600 035, India

First edition 1993

© 1993 J.S.M. Hutchinson

Typeset in 10/12 pt Palatino by Best-set Typesetter Ltd., Hong Kong
Printed in Great Britain by St Edmundsbury Press, Bury St Edmunds, Suffolk

ISBN   0 412 44310 4

A catalogue record for this book is available from the British Library

Library of Congress Cataloging-in-Publication data available

∞ Printed on permanent acid-free text paper, manufactured in accordance with the proposed ANSI/NISO Z 39.48-199X and ANSI Z 39.48-1984

To Helen

Contents

Foreword

I believe that the author has hit upon a novel idea for a different kind of book that will fill an empty niche. It is his clear intent to bring together under one cover the vast array of information available concerning reproductive biology. The book is primarily intended for use by students and teachers but it will also appeal to the intellectually curious lay public. To that end it supplies much valuable information concerning a bodily function about which their knowledge is often wanting or misinformed.

This volume deals only with the established factual information and is not burdened with the detailed documentary and supporting evidence, as is customary with books designed for active investigators. Like other fields in biology and medicine the amount of information that has been amassed in the reproductive sciences is beyond the grasp of even the most serious-minded students. Time is the limiting factor in their formal educational experience. The students' legendary lament is for 'the cold dope', or as that great English sleuth was wont to warn, 'Just give me the facts my dear Watson, just the facts'.

Endocrinology and comparative endocrinology continue to be treated as separate disciplines but the trend of the times is to abolish boundaries between long-standing disciplines and academic departments. Lines are currently being redrawn to circumscribe common scientific subject matter. In a break with tradition this volume represents an amalgamation of the reproductive biology of domestic and farm animals, humans, and wild mammals.

The neural and hormonal mechanisms controlling the normal reproductive process in both sexes is explained to the extent of present understanding. This encompasses every aspect of reproductive biology from the triggering of mating behaviour by the changing photoperiod to postpartum lactational amenorrhoea. Highlighting this account is the newly gained ability to manipulate the reproductive system. The variety of available contraceptives, other means of birth control and the loci of intervention sites are discussed. These matters are of critical importance for restraining the excessive expansion of the human population, for curbing the birth of unwanted offspring, and for overcoming the distressing plight of infertile couples.

In this age of revolutionary scientific and technological advances the reproductive sciences have startled the imagination with such spectacular achievements as *in vitro* fertilization, embryo transfers and storage, genetic engineering, cloning, and gene transfer. These and other social matters such as abortion and surrogate motherhood have raised concerns that involve legal, moral, ethical, and religious issues, all of which are well outside the scientific arena and are critically discussed.

Roy O. Greep
September, 1992

Preface

The perceived need to find efficient and acceptable methods of solving problems of overpopulation, family planning, and alleviating infertility in man, improving fertility, production, and breeding potential of domestic species, and conserving and managing wild and endangered species has led to a continued wide interest in reproductive biology and methods for its manipulation in mammals and man.

Despite the fact that mammals display a wide variation in reproductive biology there is an increased emphasis on interdisciplinary teaching and research. In addition, boundaries between specifically medical, veterinary, and ecological aspects of different subjects is diminishing, and solutions to problems are often seen to lie across boundaries. The specific needs and objectives for manipulating reproduction in animals and in man may be either very similar or very different, but the approaches and methods used depend on the integrated knowledge obtained from a wide array of different species and disciplines. This book aims to highlight the value of such an approach.

My interest in reproductive endocrinology was inspired by my teachers E.J.W. Barrington, J. Hammond, H.A. Robertson, F.T.G. Prunty and R.O. Greep, who kindly wrote the Foreword. This book is derived from lectures given over the years to biological, agricultural, and medical students. After chapters on the challenges for reproductive biology and on mammalian reproductive biology and its control, reproductive rhythms and limitations and the factors that control them are discussed. Possible points of intervention for the control of reproduction are then outlined and methods for the assessment of reproductive function essential for the understanding and interpretation of normal, abnormal and manipulated reproduction are discussed in detail. Methods for both reproductive stimulation and inhibition are discussed and attention given to the social and ethical considerations involved. At the end some suggestions are given for further reading, there is a glossary of terms and an index.

Finally, I offer special thanks to my wife, to whom this book is dedicated, for her unending encouragement and help; to my colleagues, T. Atkinson, P.J. Broadbent, K.E. Kendle, M.A. Radcliffe and B. Seddon for their helpful comments; and to generations of graduate and undergraduate students who have listened to and challenged me.

J.S. Morley Hutchinson
October, 1992

CHAPTER 1

Challenges for reproductive biology

As the world population has increased, the need for food has increased. Man has moved towards a more intensive occupation of land, and the space for wild and domestic animals has decreased. Wild species are moving into closer contact with man. The need for their conservation (or in the case of pest species, their control) has become acute and, for large mammals at least, managed herds and nature parks may become the norm. Increasing human populations mean a need for improved efficiency in the production of domestic animals and for methods for the voluntary restriction of births. Conversely, childlessness remains an acute personal problem for many. All of these factors create a requirement for the ability to manipulate reproduction. They, therefore, represent the collective challenges for reproductive biology. In this book the integrated problems related to the manipulation of reproduction in mammals including man will be discussed.

1.1 HUMAN POPULATION INCREASES AND DECREASES

At the start of the first century A.D. the population of the world was probably around 250 million. Subsequent population increase was restricted by the high death rate due to disease, war, poor nutrition, or even famine. By 1830 world population had reached about 1 thousand million (Figure 1.1). Since then, the world population has continued to increase, reaching 2 thousand million by about 1930, 3 thousand million after only another 30 years, 4 thousand million by 1975 and 5 thousand million by 1987. These increases do not reflect increases in fertility and more children being born; rather, many of the children born have survived much longer, a trend that is likely to continue (Table 1.1). In addition, until recently, spacing of births was increased by lactational amenorrhoea (see Chapter 3, page 49), which is reduced by bottle feeding. There has also been a spectacular decline in the age at puberty in the developed world; this is still falling in the less developed world, probably due to better nutrition and lack of disease.

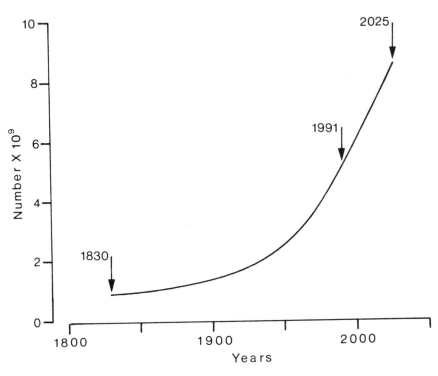

Fig. 1.1 World population 1830–1991 with a projection to 2025.

Table 1.1 Projected life expectancy in selected countries

Country	1965	1985	2005	2025
Ethiopia	37	43	51	59
China	44	67	72	75
Burma	45	55	64	70
UK	71	74	76	77
US	70	74	76	77

If the future growth of the world population does not exceed that projected by the United Nations Population Fund, the 5.4 thousand million people in the world in 1991 will increase to 7.2 thousand million in 2010 and to 8.5 thousand million in 2025. To achieve even these targets, the numbers of children born will need to fall. Such reductions in births can be achieved by a voluntary limitation: there is a wide demand for contraception which is outstripping supply. The size of families is beginning to decline worldwide: in 1991, 51% of world married couples were using contraception.

Table 1.2 Growth rates and doubling times of populations

Growth rate (%)	Doubling time (years)
1	69
2	35
3	23
4	17

The increase in population growth rate varies in different countries, being higher in much of the less developed world than in the developed world, where increases are already nearly stable (under 1%, representing a doubling in 69 years, see Table 1.2) or even falling.

Populations that are increasing fastest are not, however, the main depletors of world resources. The developed world far outstrips the others in its *per capita* use of resources and pollution potential.

The current UK population is stable at about 57 million, but some projections suggest that deaths may outstrip births by 2030, causing a decrease in population. Such decreases in births reflect changes in social habits, including delayed marriage and the desire of many women to have careers before babies. Older women are, however, less fertile than younger women leading to an increased demand for infertility treatment. Increased life expectancy coupled with a decreased birth rate leads to the challenge of an ageing population. Such challenges may be more acute in less developed countries (see Table 1.1), by producing sharp shifts in the balance of societies and an increase in welfare needs. Europe may see perhaps a doubling in the number of people over the age of 85 between 1980 and 2020, but increases in parts of South America and South-east Asia may be four-fold. This leads to an increasing dependence on the family. Careful support systems need to be devised or population targets will not be met as people insist that children are essential to provide for their old age.

1.2 EFFECTS OF OVERPOPULATION

It is a paradox that at a time when the European Economic Community, for example, is producing surplus food, and when quotas and 'set-aside' (placing land out of production) are being instituted to reduce food production, many people in the world are hungry and many suffer, at least periodically, from malnutrition and even starvation.

As, and if, world population increases it will become progressively harder, despite increased efforts, to provide an adequate food supply for all. Actions to increase food production have frequently been counterproductive due to over-exploitation of land and a loss of ecological balance, involving deforestation, monoculture, loss of grazing, soil erosion, extinction of species, and global warming due to accumulation of greenhouse gases. Longer term solutions must surely be based on a maintenance or re-establishment of balanced communities where population, food supply, agricultural systems, and the environment are in equilibrium. Such equilibria may be achieved within mixed farming systems where livestock, which may be sources of fibre, fuel, fertilizer, or traction as well as sources of high quality protein, will have a place alongside the growth of staple crops. Animals, particularly ruminants, have a special role as forages inedible to man are converted into edible animal products.

1.3 EFFICIENCY OF ANIMAL PRODUCTION

An acceptance that animal agriculture may have a continued place in the balance of world agriculture includes the need for such systems to be as efficient as possible. Methods to improve efficiency include all aspects of husbandry compatible with animal welfare, increasing genetic potential, and increasing reproductive potential. Specifically, reproductive potential can be improved either by improving normal reproduction up to the optimal potential for the particular species or breed under the prevailing environmental conditions, or by the use of novel interventions and manipulations.

1.4 WILD AND ENDANGERED SPECIES

Like man, wild mammals show a number of density-dependent characteristics which enable them to reach an equilibrium with their environment, including social organization and behaviour, food supply, reproductive patterns, disease, habitat preference and dispersal, and interactions with other species including predators and man. The increasing human population has reduced the available space, particularly for larger wild mammals, and all have progressively closer contact with man. The aim of conservation is to prevent individual species or even whole communities becoming extinct locally, regionally, or globally. Conservation of all species requires an enlightened attitude to land use and management of the habitat for their survival, linked with specific procedures for certain species such as captive breeding and reintroduction to the wild. Both present huge problems. The former requires creation of an adequate artificial environmental and social niche for success; the latter poses questions such as unknown ability of the released animal to adapt to a foreign environment, find food, cope with predators, and combat disease. Neither habitat protection nor captive breeding alone is likely to solve the problem of large species conservation. Existing plans for the captive breeding of about 200 species by the year 2000 represents only a small fraction of the overall need. Genetic resource banks of frozen oocytes, embryos, and sperm may, however, not only offer an insurance against the total loss of a species and the ability to maintain genetic diversity in both wild and captive populations, but also facilitate captive breeding programmes by allowing easy transport to remote areas, use over many generations, and the removal of male preference and sexual incompatability. Nature parks and managed herds have become more necessary for the survival of many species. In such environments wild species can lead relatively undisturbed lives, albeit under pressures from tourism, recreation, or as human food animals.

1.5 WILD VS. DOMESTIC ANIMALS

Domestic livestock are derived from wild species, but fundamental differences exist. Domestic animals have been selectively bred to provide meat, milk, or fibre and are usually managed to ensure their wellbeing, man providing food, protection, and healthcare. In contrast, wild species have adapted to their environments by natural selection and are 'better able' to survive alone than domestic animals. Domestication, like captivity, has affected all aspects of reproduction, perhaps partly due to influences of stress and nutrition. Evidence from pigs, cattle and sheep, for example, suggests that domestic breeds show less seasonality in their breeding. The boundary between wild and domestic/farm species is being lessened. The category of domestic animals includes all species whose breeding is or can be controlled by man. These will usually not include wild species that have been tamed. However, taming, that is taking animals from the wild environment and applying food and shelter, merges into domestication. On the other hand, domestic animals may be released into the wild and become a so-called feral species.

The balance of wild and endangered species is greatly influenced by the presence or absence of predators. As species become progressively protected by man in reserves and parks their communities become more artificial and they may require more external management. Man is the greatest threat to endangered species through removing habitats, farming, disturbance by traffic, aeroplanes, even tourists and poaching. Particularly with the larger mammals the question of survival may be in doubt, without some form of adaptation.

Some species that are successfully adapted to life near man may become pests. Such species may have become more naturally successful or may have been introduced artificially (rabbits in Australia), inadvertently (escaped fur species like mink and coypu) or foolishly (abandoned dogs and cats). Management and control is desirable, not only for pest populations but also for the highly desired species whose numbers may have increased under protected conditions where natural control by predators may be absent, and where wide natural fluctuations in number are not desirable or practicable. It may be necessary to reorganize some species of animals into new populations, some in captivity, some in the wild, with the animals being managed as a whole as a megazoo. Such management of wild species is controversial, but for some of the larger species of mammals at least it may be the only way to avoid extinction.

1.6 REPRODUCTIVE MANAGEMENT AND CONTROL

The manipulation of reproduction may take a wide variety of forms depending on the particular aims and circumstances, and may involve the restoration of normal reproductive activity, the stimulation of increased activity, or the inhibition of activity. The general and specific aims for each circumstance must be clearly identified and the appropriate financial, social, and ethical costs and the cost-effectiveness must be evaluated. Methods for stimulation of reproduction function in farm animals or for individual family planning may need to be cost-effective and essentially 100% efficient to be acceptable; less cost-effective and less efficient procedures may be acceptable to help conserve a highly endangered species when no other means is available. Any process of reproductive management or control of farm animals, wild or endangered mammals, or man must also be viewed in a wider context of general management, health, and welfare. Any procedure will be influenced by the nutritive state, disease, the environment, the stage of reproduction at onset of treatment, age, ecology, and by social and economic factors, depending on the species. In the wider context of acceptability, concern for animal and human welfare and rights and the wider interest in 'green issues' bring with them increased public awareness of wild life conservation and human population problems. There are suspicions about the use of drugs and the application of molecular biological techniques, and a demand for more natural, biological control procedures where possible.

CHAPTER 2

Background reproductive biology

2.1 GENERAL FUNCTIONS AND CONTROL OF REPRODUCTION (Figure 2.1)

The primary reproductive organs, the gonads (ovaries and testes), produce both the gametes (oocytes and sperm) and an array of hormones. In mammals, sperm are transferred to the female tract at mating. Fertilization takes place and, other than in monotremes (e.g. anteater and platypus), pregnancy follows. Nurturing of the early offspring within the mother is followed at parturition by the birth of live young. The hormones produced by the gonads affect reproductive behaviour, the development and maintenance of secondary sex characteristics and accessory glands and structures, and, along with other hormones, particularly those from the anterior pituitary, the integrated control of gonadal function. Reproduction is also controlled by influences from and interaction with the external social environment, including individuals of the same and other species, and the physical environment, including light, temperature, and nutrition, acting through stimulation or inhibition of the brain, hypothalamus, and anterior pituitary.

2.2 DIVERSITY OF REPRODUCTION IN MAMMALS

Although much is known about reproductive processes in domestic species, laboratory species, and man, very little is known about even basic mechanisms in other species, let alone the detailed information needed to manipulate their reproduction effectively. There is no standard pattern of reproduction in mammals: there is great diversity, perhaps more than any other physiological parameter, sometimes even between apparently closely related species.

Mammalian species show marked differences in the time of onset of puberty, responses to the environment (e.g. seasonal and non-seasonal breeders), social interactions and behaviour, the number of

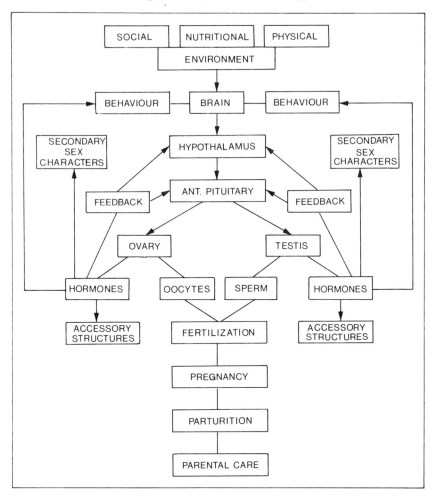

Fig. 2.1 General aspects of reproductive function and control.

ovulations (single, twins, litters), and the factors controlling ovulation (spontaneous or induced by coitus). All such differences may represent adaptations to particular ecological, environmental, and social systems which enable the young to be born and weaned at a time of favourable temperature and food supply. The monotremes differ from other mammals in that the posterior end of the intestine, the ducts of the excretory systems and the genital ducts form a common cloaca with a common opening. These animals lay shell-covered eggs that are incubated and hatched outside the body of the mother. The reproduction of marsupials differs from that of other mammals in that most females have an abdominal pouch within which the young are carried.

They lack a compete placenta. Instead, a yolk-sac placenta is formed through which uterine secretions are absorbed, although in some species a more developed placenta is formed. The gestation period is short, compared with eutherian mammals of equivalent size, and the young are born practically in the embryonic state.

Perhaps the most variable component of eutherian mammal reproduction is the adaptation by which pregnancy evolved to allow viviparity. Such adaptations took place late in evolutionary history and have taken place in a variety of ways. All eutherian mammals initially maintain pregnancy by an extension of the luteal phase (see page 40) but the methods of pregnancy, the methods of corpus luteal control, the subsequent source of progesterone (luteal or placental), the mode of implantation and placentation, and the number of offspring vary enormously.

In general there are two groups of eutherian mammals. The first produces large litters of poorly developed offspring, born after a short gestation period (insectivores, carnivores and rodents, though not hystrichomorphs like the guinea-pig). The second group produces small litters of well-developed offspring, born after a long period of gestation (artiodactyls, perissodactyls, pinnepedes, primates, cetaceans).

Differences also exist between different species and sexes in the time spent on different activities, including reproduction. In man certain individuals and groups may have a 'harder' life than others, and every effort is placed into obtaining food. The production of many children may be regarded as an insurance policy. The majority of wild species spend most of their waking hours looking for food (carnivores) or feeding (herbivores), and much of the rest of the time sleeping or hibernating. The rest of their time is spent seeking mates and 'seeing off' rivals; or, in females, in pregnancy and lactation. Domestic species have food, housing, and protection provided: 'artificial' breeding and mating programmes may be the norm. Mate selection and male rivalry have been eliminated, and where artificial insemination is used the males have become simply sperm producers. Mammals in wildlife parks or zoos may have become adapted to an impaired or modified lifestyle. Some species, such as rats, foxes, and racoons, have adapted to urban living and may be even more successful there than in the wild. It may, however, be essential to understand the social structure of a species (e.g. males, subordinates, female–female interaction, monogamous pairings) to appreciate the possibilities for adaptation to a modified habitat or to close association with man. Carnivores may require a modified food presentation, and various forms of behavioural enrichment may be necessary to produce daily routines to mimic those prevailing in a natural ecosystem.

An understanding of the interactions between members of a species is important, therefore, to be able to manipulate its reproduction successfully. For different situations, such as human contraception or assisted conception, the induction of out-of-season breeding, or gene conservation of a wild/endangered species, very different objectives for manipulation may prevail. The overall principle of reproductive manipulation, regardless of the specific objectives, will be, however, the targeting of a particular parameter of reproduction in a specific, predictable, safe, reversible, and ethically acceptable way. A general outline of mammalian reproduction is, therefore, appropriate to highlight such key targets of reproduction control.

2.3 HORMONES AND REPRODUCTION

Normal reproductive functions, including the function of the ovaries and testes, are controlled by an interaction of pheromonal, nervous, hormonal, and local chemical mechanisms. In considering the integrated system involved, however, such distinctions seem artificial. This is particularly true for the restriction of the term 'hormone' to a signalling chemical secreted from a ductless (endocrine) gland into the bloodstream for function elsewhere (Figure 2.2). Here, the term will be used in a wider more general context. The different types of system involved are illustrated in Figure 2.2: a system is defined by whether the origin of a signalling chemical is from a nerve cell or a non-nerve cell, by the mode of transmission of the signal and by the type of target. In neurocrine control a signalling chemical produced in a nerve cell is released at the nerve ending to affect its target (e.g. innervation of the ovary). In neuroendocrine control, the product of a nerve cell is released at nerve endings into the bloodstream to act elsewhere. Gonadotrophin-releasing hormone control of gonadotrophin synthesis and release is such a system (see page 23). In a classical endocrine system the hormone is secreted by cells of a gland for transportation *via* the blood to function elsewhere (e.g. anterior pituitary gonadotrophins affecting the gonad). Paracrine control involves the product of the cell acting locally (e.g. within the ovary); in autocrine control, the action is on the cell of origin; and in intracrine control the action is within the cell of origin without secretion. In socioendocrine control a so-called pheromone is produced in one individual and transferred *via* the external environment to affect another individual. In reproductive control such mechanisms react and interact in complex ways; some 'hormones' act in more than one way.

2.3.1 Naturally occurring hormones

The main endogenous hormones and related chemicals involved in reproduction are shown in Table 2.1. Their detailed origins and actions

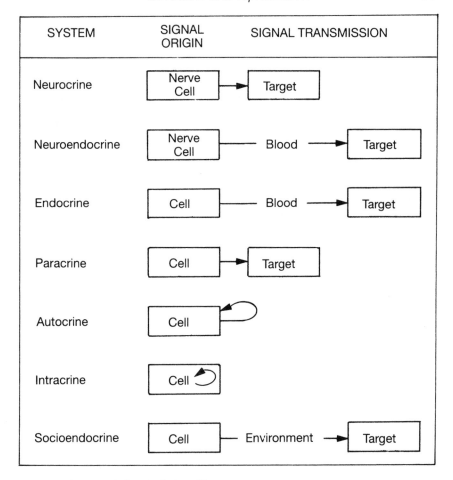

Fig. 2.2 Types of chemical signalling system.

will be discussed later; their general nature and structure (Figures 2.3–2.7) are discussed here.

Melatonin, secreted by the pineal gland, is an indoleamine (Figure 2.3a), whereas the other neurotransmitters are either amines (e.g. 5-hydroxytryptamine, Figure 2.3b; γ-amino butyric acid) or catecholamines with the two hydroxyl groups on the benzene ring (e.g. dopamine; noradrenaline, Figure 2.3c). The opiate peptides are encephalins and endorphins. The gonadotrophin-releasing hormone (GnRH) of the hypothalamus is a decapeptide (Figure 2.3d). GnRH is derived from a larger protein, prepro-GnRH, which contains GnRH and other sequences including GnRH-associated peptide (GAP). The gonadotrophins synthesized by the anterior pituitary (follicle-stimulating hormone, FSH; luteinizing hormone, LH) are glycoproteins, consisting of

Table 2.1 Some natural hormones involved in the control of reproduction

Hormones and related substances	Class of substance
Brain hormones and some neurotransmitters	
Melatonin	Indoleamine
Dopamine	Catecholamine
Noradrenaline	Catecholamine
5-Hydroxytryptamine	Amine
γ-Amino butyric acid	Amine
Opiate peptides	Peptide
Hypothalamus	
Oxytocin	Peptide
Gonadotrophin-releasing hormone (GnRH)	Peptide
Prolactin-releasing and -inhibiting factors	Various
Anterior pituitary	
Gonadotrophins	
Follicle-stimulating hormone (FSH)	Glycoprotein
Luteinizing hormone (LH)	Glycoprotein
Prolactin	Protein
Gonadal/Placental steroids	
Androgens (e.g. testosterone)	Steroid
Oestrogens (e.g. oestradiol-17β, oestrone, oestriol)	Steroid
Progestogens (e.g. progesterone, 17-hydroxyprogesterone)	Steroid
Gonadal peptides	
Inhibins	Peptide
Activins	Peptide
Relaxin	Peptide
Oxytocin	Peptide
Growth 'factors' (e.g. Transforming growth factors, TGFs, Insulin-like growth factor I, IGF-I)	Peptide
Conceptus signalling substances	
Chorionic gonadotrophins (e.g. eCG, hCG)	Glycoproteins
Trophoblast proteins (e.g. oTP, bTP)	Interferon-like
Uterine luteolysins	
Prostaglandins (e.g. $PGF_{2\alpha}$)	Lipid-soluble acids
Other systems	
Other hormones associated with reproduction (e.g. Growth hormone, Adrenal and Thyroid hormones)	Various

an α and β peptide chain (in which the α chain is essentially species-specific and the β chain is hormone-specific) linked to a carbohydrate moiety, which affects the half-life of the molecule in the blood. Prolactin from the anterior pituitary is a protein. Its secretion is controlled by an

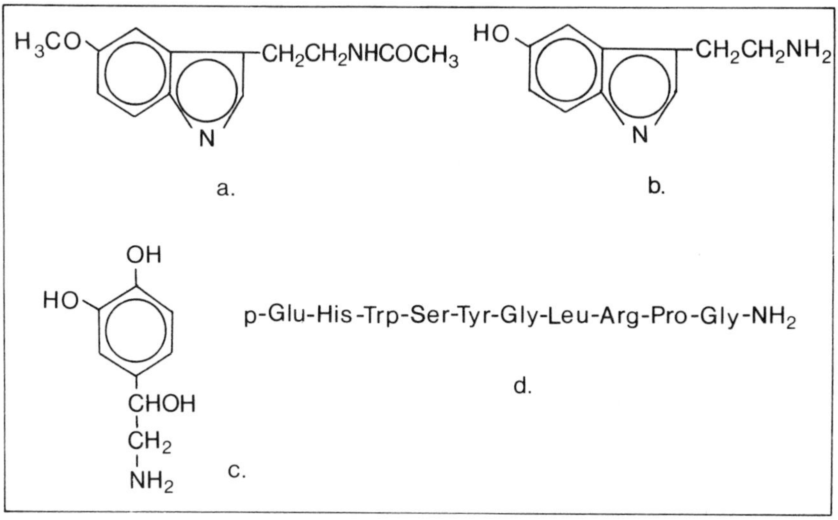

Fig. 2.3 The structures of (a) melatonin, (b) 5-hydroxytryptamine, (c) noradrenaline and (d) gonadotrophin-releasing hormone.

inhibitory action of the hypothalamus; this probably involves dopamine, but other inhibitory and stimulatory factors have been described.

All steroid hormones have the general structural formula shown in Figure 2.4a. The numbering of the carbon atoms is shown in Figure 2.4b. All naturally occurring progestogens (e.g. progesterone, Figure 2.5a) and corticosteroids have 21 carbon atoms (C21). Androgens (e.g. testosterone, Figure 2.5b) have 19 carbon atoms (C19) and oestrogens (e.g. oestradiol-17β, Figure 2.5c) have 18 carbon atoms (C18). The biosynthetic pathways of all steroid-producing tissues are very similar (Figure 2.6). In essence all steroid-synthesizing tissues have 'potentially' the same set of enzymes, the specific pathways followed and the specific hormones produced depending on the specific enzymes present in a particular tissue at a particular time. Only one tissue type may be involved, as in the production of testosterone by testicular Leydig cells, or more than one tissue may be involved, as in the ovarian follicle, where androgens produced by theca cells can be transferred to granulosa cells for conversion to oestrogens (see Figure 2.21, page 33).

The inhibins, activins (Figure 2.7a), and follistatins are three families of polypeptides that modulate pituitary FSH release and act locally within the gonads. The inhibins inhibit FSH release and are composed of a common α subunit bonded to either a βA subunit (inhibin A) or a βB subunit (inhibin B). The activins, which stimulate FSH release, are composed of either two βA subunits of inhibin A to form activin A or the βA and βB subunits of inhibins A and B to form activin AB. Follistatins are glycosylated proteins present in at least three different

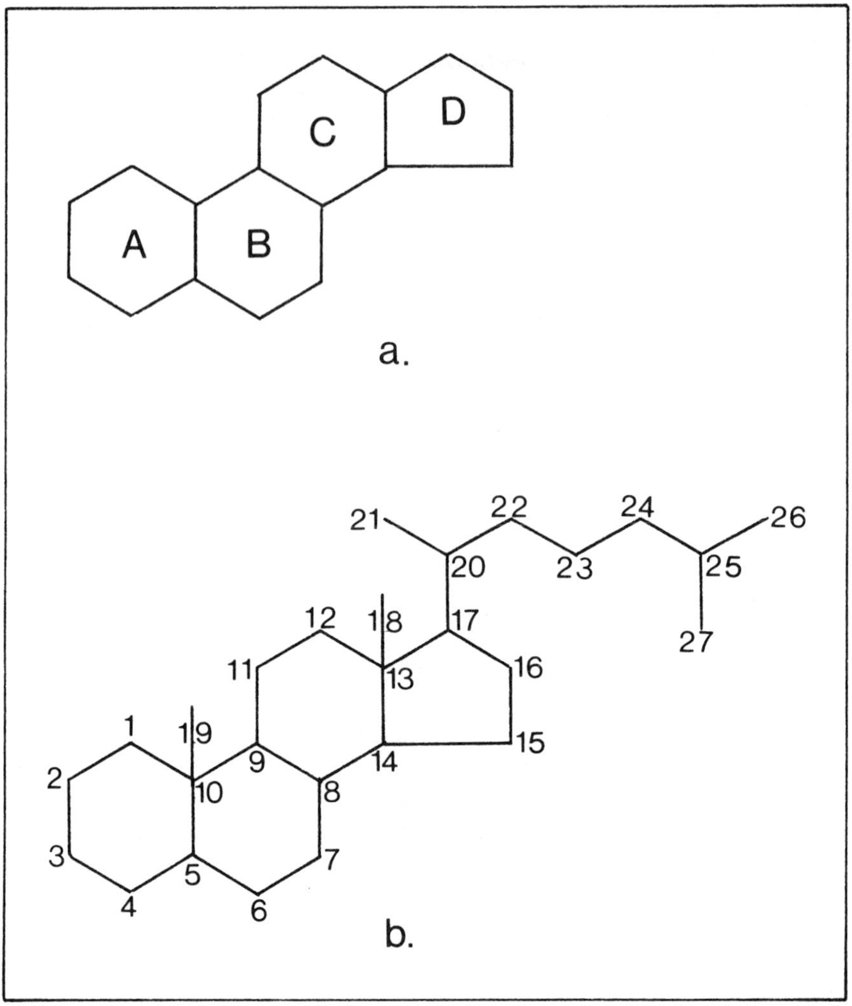

Fig. 2.4 The general structure of steroids. (a) The common ring structure of all steroids consists of three six-membered carbon rings and one five-membered carbon ring, in the configuration shown, and named A–D, respectively. (b) The numbering of the carbon atoms in the steroids and precursor molecules (e.g. cholesterol).

molecular forms. All usually inhibit FSH release, but they may be more important as binding proteins for activins. An N-terminal portion of inhibin called α-N peptide possesses no inhibin-like action but may act locally to facilitate the process of ovulation. Many of the other locally and peripherally acting hormones of the gonads (relaxin, oxytocin and the so-called growth factors) are peptide in nature. The specific conceptus-signalling substances are either glycoprotein hormone-like

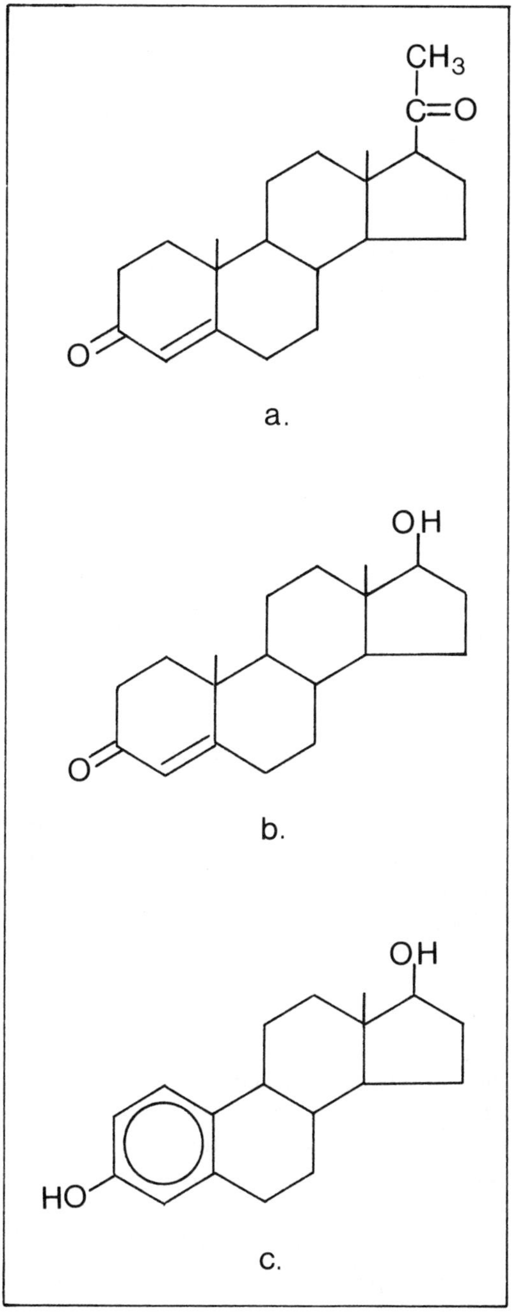

Fig. 2.5 The structures of three naturally occurring steroids: (a) progesterone, (b) testosterone and (c) oestradiol-17β.

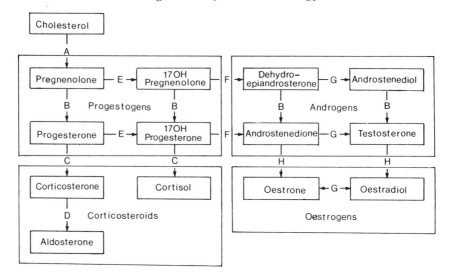

Fig. 2.6 The general pathway of steroid biosynthesis. The enzyme steps include A. cholesterol side-chain cleavage; B. 3β-hydroxysteroid dehydrogenase/ Δ^{5-4}-isomerase; C. 21- and 11β-hydroxylase; D. 18-hydroxylase and 18-hydroxysteroid dehydrogenase; E. 17-hydroxylase; F. C-17, 20-lyase; G. 17β-hydroxysteroid dehydrogenase; H. aromatase.

(chorionic gonadotrophins) or interferon-like (trophoblast proteins). Prostaglandin $F_{2\alpha}$ is one of a group of lipid acids (Figure 2.7b).

Neuroendocrine and endocrine hormones are normally secreted into the bloodstream in a pulsatile manner. The size and frequency of the pulses depend on the rate of the stimulus and the physiological state of the gland (see Figure 5.5, page 81). The blood concentration of the hormone reflects not only its rate of production by the gland but also its rate of conversion (metabolism) or clearance from the bloodsteam. The hormone may be bound to protein for transport in the blood (e.g. sex hormone-binding globulin): the 'free' unbound fraction probably represents the biologically active fraction. Some hormones may exist in different forms (e.g. gonadotrophins) which have different potencies and biological half-lives (see also Chapter 4, page 62). Some classes of hormone (e.g. oestrogens, androgens) may be present in the bloodstream as more than one substance (e.g. oestrogens as oestrone, oestradiol-17β, oestriol) due to their differential synthesis, conversion, or metabolism. Such different oestrogens may have different biological actions.

2.3.2 Hormone action

Many hormones exert their actions *via* specific receptors, either on the cell plasma membrane (e.g. protein/peptides) or within the cell (e.g.

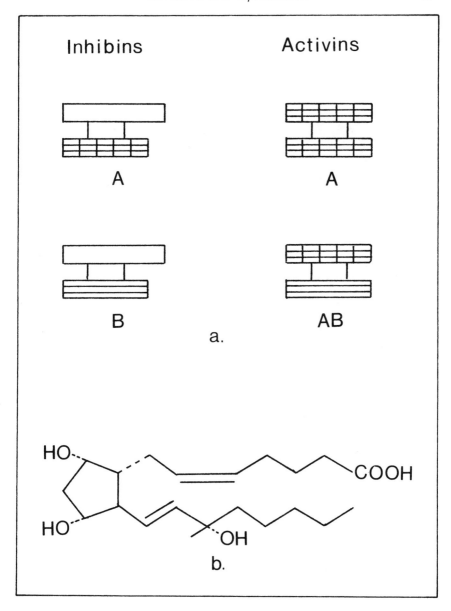

Fig. 2.7 The structure of (a) the inhibins, the activins and (b) prostaglandin $F_{2\alpha}$.

steroids) (Figure 2.8). The binding of a hormone to its receptor leads to a cascade of reactions leading to the specific response within the cell. The action of a hormone will, therefore, not only reflect the presence of active hormone at an appropriate concentration but also the presence of its receptor at the target site. Not only can receptors be 'stimulated'

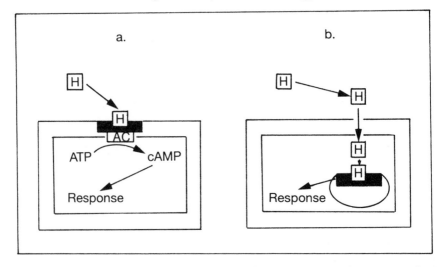

Fig. 2.8 The action of hormones *via* specific receptors. (a) One action of a protein hormone (H, e.g. a gonadotrophin) *via* a cell membrane receptor (shaded) and a second messenger, cyclic 3′5′-adenosine monophosphate (cAMP), formed from ATP, involving a G protein receptor linked adenylate cyclase (AC). (b) Action of a steroid hormone (H, e.g. an oestrogen) *via* an intracellular nuclear receptor (shaded).

by the presence of the appropriate hormone but they may sometimes be blocked and their numbers reduced. A substance that binds to a hormone receptor but fails to direct the appropriate cellular response is an antihormone. Such a substance acting on an oestrogen or progesterone receptor will be termed an antioestrogen or antiprogestogen, respectively. The receptors on a given target cell are subject to change and may be 'primed' by hormone action. Thus, priming by oestradiol and GnRH may determine the response of the anterior pituitary to GnRH. Receptors may also 'down-regulate': prolonged action of a hormone on a receptor may lower rather than raise the effect. Thus, GnRH stimulates LH release from the anterior pituitary if given episodically but inhibits LH release if given continuously. The effect of a given hormone may also be modified by its metabolism within the target cell. Thus androgens may be metabolized to oestrogens, testosterone reduced to dihydrotestosterone, and oestrogens hydroxylated to form catecholoestrogens.

2.3.3 Hormone feedback

Hormonal control mechanisms operate in an integrated fashion involving the phenomenon termed 'feedback'. The basic mechanisms involved in negative and positive feedback are shown in Figure 2.9. A

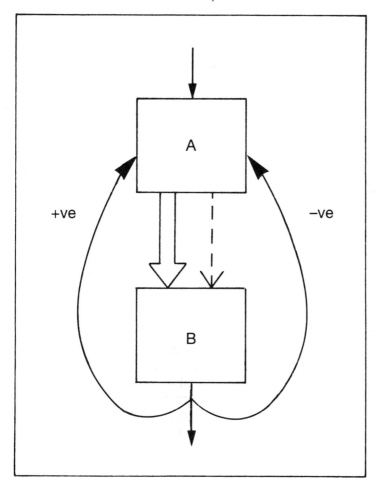

Fig. 2.9 The basic mechanisms of negative and positive feedback (for detailed description, see text).

mechanism by which the product of gland A stimulates a second gland B to produce product B, and B then acts to control A by decreasing the production of A, is called negative feedback. If, however, B acts back to control A by increasing the production of A, this is termed positive feedback. Negative feedback tends to lead to an episodic but even level of function, whereas positive feedback leads to a surge or cascade of activity. Such a model as applied to the control of the gonads is shown in Figure 2.10. The gonads are controlled by the sequential action of extra-hypothalamic factors on the hypothalamus, hypothalamic GnRH on the anterior pituitary, and anterior pituitary gonadotrophins on the gonads. The products of the gonads (steroids and peptides) act *via* the bloodstream to control this cascade in either a stimulatory (positive

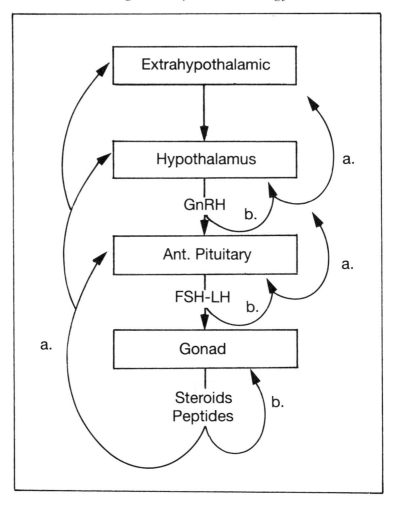

Fig. 2.10 The principles of feedback as they apply to the control of the hypothalamo–pituitary–gonadal system. (a) Long-loop feedback, (b) short-loop feedback involving paracrine or autocrine mechanisms.

feedback) or an inhibitory (negative feedback) manner at the level of either the anterior pituitary, to affect gonadotrophin production, and/or the anterior pituitary response to GnRH, or the hypothalamus and/or the extra hypothalamic mechanisms controlling it. Such feedback is often termed long-loop feedback. Other feedback loops may also operate (Figure 2.10). Ovarian products may control ovarian function directly, gonadotrophins may control gonadotrophin production by acting on either GnRH or gonadotrophin production or release, and GnRH may act to control its own synthesis or release.

Such mechanisms operating in either an autocrine or paracrine way (see page 11) may be called short-loop feedback.

2.3.4 The hormonal control of reproduction

Optimal reproductive function involves functional interactions between the environment, the brain, the hypothalamus, the anterior pituitary

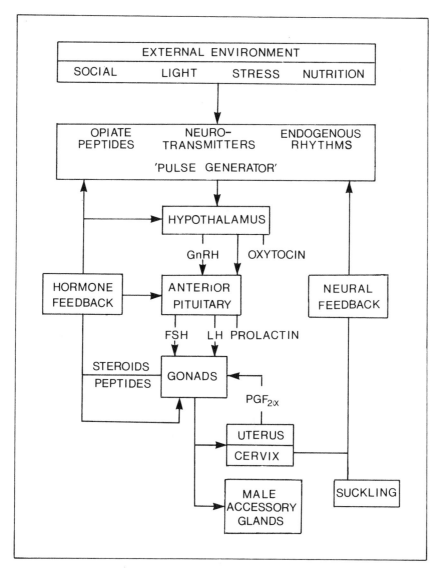

Fig. 2.11 Endogenous and exogenous factors controlling reproduction.

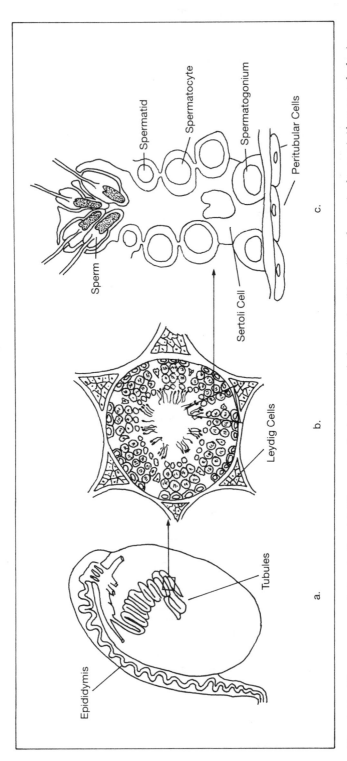

Fig. 2.12 The stucture of the testis. (a) The general structure of the testis is diagramatically drawn in a testis with the capsule removed; (ii) the epididymis is a greatly coiled structure). (b) The general structure of a seminiferous tubule in cross-section. (c) The detailed structure of the spermatogenic elements and their relationship to the Sertoli cell. [*After*: Griffin, J.E., Wilson, J.D. (1985) in *William's Endocrinology*, 7th Edn. (eds J.D. Wilson, D.W. Foster) Philadelphia, Saunders, p. 262; Skinner, M.K. (1991) *Endocrine Rev.*, **12**, 46].

gland, the gonads, and the accessory organs (Figure 2.11). The gonadotrophins, FSH and LH, are involved in the production of both gametes and hormones by the gonads. Prolactin may also be involved in certain gonadal functions in some species. The synthesis and release of gonadotrophins is directly controlled by GnRH; this is synthesized in neurones of the hypothalamus, then released at nerve endings in the median eminence, to stimulate the anterior pituitary, *via* the hypophysial-portal blood system. The synthesis and episodic release of GnRH is controlled by a complex integrated system, the 'pulse generator', which may involve various neurotransmitters, opiate peptides, melatonin, a variety of hormonal and neural feedback loops, and endogenous hypothalamic and brain rhythms to control, *via* the gonadotrophins, specific phases of gonadal function in the male and the female. The complex brain–hypothalamus–pituitary complex is also influenced by inputs from the external environment, such as light, food supply, temperature, suckling, social cues, and stress, and the gonad is also controlled by a complex of internal autocrine and paracrine mechanisms.

2.4   THE TESTIS (adult, Figure 2.12)

The testis consists of seminiferous tubules (which contain both the spermatogenic elements, and the Sertoli cells which play a role in support, nutrition and hormonal integration), the peritubular cells surrounding the tubules, and the Leydig cells which lie between the seminiferous tubules and produce testosterone. Spermatogenesis is the sum of events leading to the production of spermatozoa. Stem cell spermatogonia at the base of seminiferous tubules divide (mitosis) to both maintain their number and to cyclically produce primary spermatocytes. These undergo a meiotic reduction division to form secondary spermatocytes which then undergo the second meiotic division. Four spermatids are thus formed from each primary spermatocyte, each with half the chromosomal number (Figure 2.13). The spermatids differentiate into spermatozoa. The duration of spermatogenesis is many days (Table 2.2).

Table 2.2   The duration of spermatogenesis

Species	Time (days)
Bull	61
Ram	47
Boar	39
Man	74

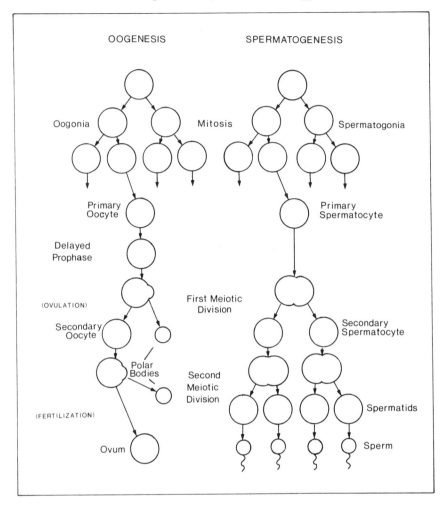

Fig. 2.13 A comparison of the events occurring during oogenesis and spermatogenesis.

Fig. 2.14 Stages of the spermatogenic cycle. Features of the spermatogenic cycle (b) are illustrated by comparison with a 4-year undergraduate university career model (a). Note that the number of stages will vary between species (e.g. 6 in man) and because of cell division, for each spermatogonium entering spermatogenesis four sperm will be produced (see Figure 2.13), whereas each student entering a university year will yield only one, even without dropouts. (*After*: Johnson L. (1990) *in Gamete Physiology* (eds R.H. Asch, J.P. Balmaceda, I. Johnston) Norwell, Serono Symposia, p. 5).

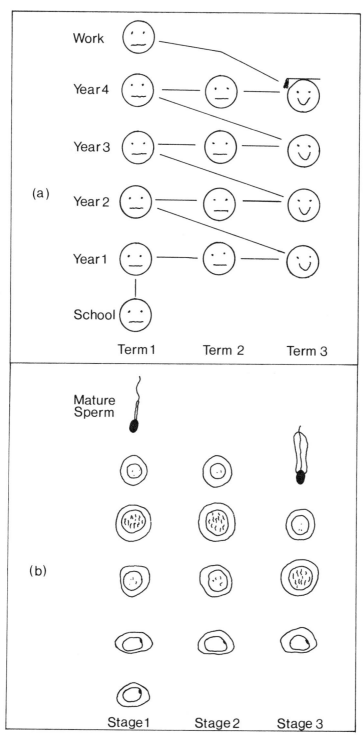

Integrated into the spermatogenic progression from spermatogonia to sperm are the so-called spermatogenic cycle and spermatogenic wave. The former is a phenomenon occurring in time, whilst the latter occurs in space. The spermatogenic cycle is the series of changes taking place at any portion of the tubule between two appearances of the same stage of development. Johnson has illustrated many aspects of the spermatogenic cycle by comparing it with a 4-year university career (see Figure 2.14). In the undergraduate student model the cycle is all the events between student entry and graduation. The two processes are similar in that the time between consecutive releases is less than the length of the entire process and the processes occur at a set rate so that students or spermatogenic cells at specific stages (e.g. term 1 in different years and cell stage 1) are always associated with students or cells at the same developmental stage. The duration of spermato-genesis is 4–5 times the cycle length, so that 4–5 germ cells at different stages of development are present in different histological views of the tubule. The number of stages of spermatogenesis (shown as only three in the model) varies between species, with 14 in the rat, eight in the horse, ram, and boar, and six in man. The spermatogenic wave is the spatial sequential order of stages along the length of the seminiferous tubule at any given time.

2.4.1 Control of testicular function

A summary of the hypothalamo–pituitary feedback control of the testis in an adult mammal is shown in Figure 2.15. Leydig cells respond

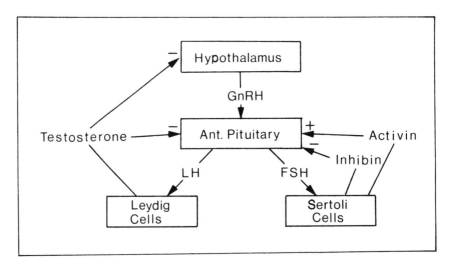

Fig. 2.15 Hypothalamus, pituitary and feedback in the control of testicular function.

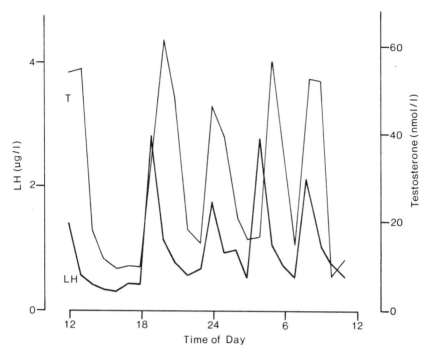

Fig. 2.16 Fluctuations in plasma LH and testosterone concentrations in a bull. (*Note*: Each episode of LH secretion is followed by a stimulus of testosterone secretion). (*From*: Katongole, C.B., Naftolin, F., Short, R.V. (1971) *J. Endocrinol.*, **50**, 460).

to episodic LH stimulation (and in some cases also to prolactin) by increased androgen production. Such androgens pass into the lymph and peripheral circulation to control the activities of the epididymis and accessory glands (e.g. seminal vesicles and prostate), secondary sex characters (e.g. breaking voice in man, antlers in deer), behaviour, libido, and pheromone production, and control LH secretion by negative feedback action. The role of such negative feedback in the physiological control of LH fluctuations (Figure 2.16) may, however, be minimal.

In addition to their peripheral actions, androgens have local actions on Sertoli cells to affect spermatogenesis either directly or *via* an action on peritubular cells to promote the production of a protein modulating Sertoli cell function (P Mod S) which subsequently acts on the Sertoli cells (Figure 2.17). Sertoli cells also respond to FSH. Such stimulation may be essential for the initiation of spermatogenesis. The interaction of FSH and androgens and their exact role in spermatogenesis is not clear, but knowledge of this interaction is clearly crucial for the development of reproduction control measures. Sertoli cells also produce a

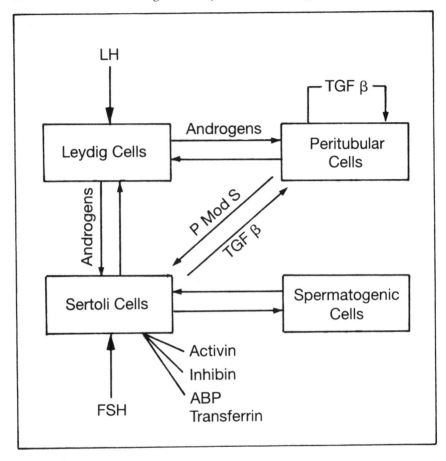

Fig. 2.17 Possible inter-relationships between Leydig, Sertoli, peritubular, and spermatogenic cells in the control of testicular function. (*From*: Skinner, M.K. (1991) *Endocrine Rev.*, **12**, 62).

number of other products, including inhibin, activin, transferrin and androgen-binding protein (ABP). Inhibins and activins are involved in negative and positive feedback control of FSH secretion respectively (Figure 2.15). This presumably takes place at the level of the pituitary to ensure specificity, as action at the hypothalamic level to affect GnRH would be non-specific and would also affect LH. The exact physio-logical roles of inhibins and activins are not known, but activin stimu-lation of FSH release may be part of the initiation of spermatogenesis at puberty or after quiescence in seasonally breeding species.

Androgen-binding protein may not be involved in the transport of androgen to spermatogenic cells, but may play a role in the transport of androgens throughout the male tract. Numerous other secretory products are produced by the components of the testis including growth factors (e.g. P Mod S, TGFβ). These substances are probably involved in the autocrine and paracrine control and reciprocal inter-action between peritubular and Leydig cells, peritubular and Sertoli cells, Sertoli cells and Leydig cells, and spermatogenic and Sertoli cells (Figure 2.17).

2.4.2 Sperm storage and transport (Figure 2.18)

After their production, sperm are passed into the epididymis for storage and maturation; from there they are ejaculated. Thus, though daily sperm output reflects the number of sperm produced by the testes, the number of sperm in the ejaculate is influenced by the number of sperm stored in the epididymis, which depends on the ejaculation frequency. The accessory glands may vary greatly between species and may include ampullary glands, seminal vesicles, prostate, bulbourethral (Cowper's) and urethral glands. During the passage of sperm down the vas deferens complex secretions are added from the accessory glands,

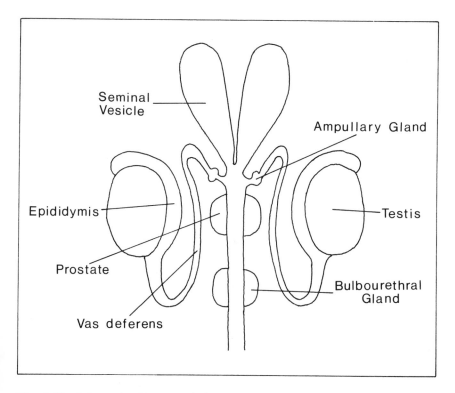

Fig. 2.18 Schematic diagram of the male reproductive tract and accessory glands.

including substances which maintain the osmolarity and pH, and which provide an energy source for the sperm (Figure 2.18).

2.5 THE OVARY

Oogonia are formed in the ovary during the embryonic period and undergo mitotic division. Unlike in spermatogenesis, however (see Figure 2.13), this mitotic phase ceases during the perinatal period, varying a little between species. Oogonia form primary oocytes which commence meiotic division and become surrounded by a layer of cells. At the dictyate stage of the prophase, meiosis is arrested. Oogonia not completing this process are lost and no further oogonia are formed throughout the life of the individual. The oocytes, surrounded by a single layer of cells, form the primordial follicular pool (Figure 2.19). Some follicles leave this pool in a continuous sequence and start to grow progressively. The oocyte grows and becomes surrounded by an acellular layer, the zona pellucida. The surrounding granulosa cells divide and the follicle enlarges, with layers of theca cells being laid down around the granulosa cells. During subsequent development the granulosa cells continue to divide and the theca cells differentiate into two layers, a vascular theca interna layer within a theca externa. Fluid begins to accumulate between the granulosa cells and the spaces join to form a fluid-filled antrum; the oocyte comes to lie on the side of the antrum, surrounded by layers of granulosa cells. Evidence from species such as sheep and cattle suggests that, starting prepubertally, cohorts of antral follicles develop in waves (see Chapter 3, page 45). Almost all follicles that start a progressive growth undergo a regressive process termed atresia. Only a few such follicles progress through to a phase of preovulatory follicular development, culminating in the process of ovulation, when the oocyte is shed. In species which usually have a single ovulation, one dominant follicle may progress to ovulation or a single dominant follicle may be actively or passively selected from a preovulatory cohort of follicles. In species which produce litters and which have multiple ovulations, a preovulatory cohort of follicles, at perhaps slightly different stages of development, progress to ovulation. At ovulation (in most species) the oocyte completes the first meiotic (reduction) division (Figure 2.13) with almost all the cytoplasm remaining with one set of chromosomes and essentially none with the other. Thus the secondary oocyte and a so-called polar body are formed. After ovulation the granulosa cells (and in some species also theca cells) remaining in the follicle differentiate into luteal cells (the process of luteinization), which grow, become vascularized and form the corpus luteum. This will be maintained for a variable period (both morphologically and functionally) depending on the species and whether

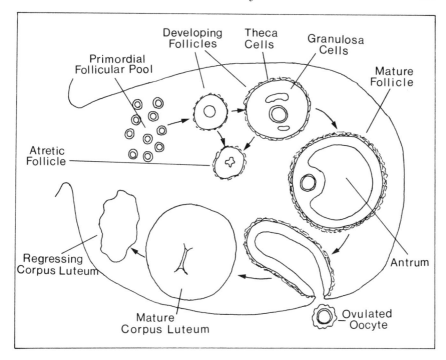

Fig. 2.19 A diagramatic cross-section of the ovary. This diagram is not intended to illustrate the appearance of the ovary at a specific time or stage of development, but rather to illustrate various sequential stages of follicular maturation, atresia, ovulation, corpus luteal development, and regression.

developing early embryos are present (see below), after which it will eventually regress.

2.5.1 Control of ovarian function

The regulatory mechanisms involved in the different stages of follicular development and atresia are varied and complex. There are two levels of control: (i) intraovarian autocrine and paracrine (both within follicles and between follicles and other ovarian components) systems involving a variety of steroid, peptide and other mechanisms; and (ii) stimulation by gonadotrophins, and feedback signals from the ovary to gonado-trophin control systems (Figure 2.20). Little is known of the factors responsible for establishing the primordial follicle pool or of those regulating the initial growth of progressive follicles from this follicular pool. The latter may be under paracrine control by inhibitory sub-stances produced by other follicles in the pool. The control of the further growth of a follicle may be divided into gonadotrophin-independent and -dependent phases. During the preantral growth

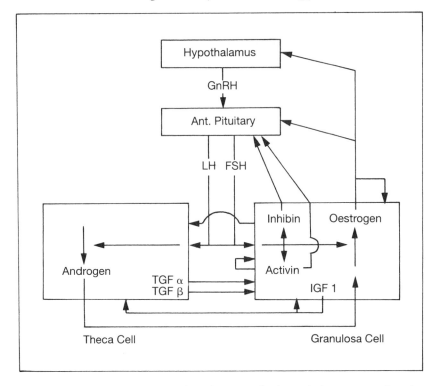

Fig. 2.20 Possible interrelationships between the hypothalamus, anterior pituitary, theca, and granulosa cells in the control of folliculogenesis. Follicular development can be divided into a gonadotrophin-independent phase controlled by intraovarian factors, and a gonadotrophin-dependent phase involving episodic hypothalamic GnRH stimulation of gonadotrophin release, the action of LH on theca cells, and the action of FSH (and LH) on granulosa cells, in association with a complex interaction with intraovarian actions of peptides and steroids and their extraovarian feedback effects on the anterior pituitary and hypothalamus. (*After*: Montgomery, G.W., McNatty, K.P., Davis, G.H. (1992) *Endocrine Rev.*, **13**, 321).

phase (and perhaps extending into the early antral stage) TGFα (or epidermal growth factor) from thecal cells probably stimulates proliferation of granulosa cells but inhibits their differentiation. Further follicular development comes under increasing control by pituitary gonadotrophins. There is evidence that both FSH and LH are required to stimulate this further follicular development and oestrogen secretion. Early antral follicles sensitive to gonadotrophin stimulation have FSH receptors on granulosa cells and LH receptors on theca cells. These theca cells produce androgens (in response to LH) which are transferred to granulosa cells, which under FSH stimulation convert androgens to oestrogens, the so-called two cell, two hormone hypothesis

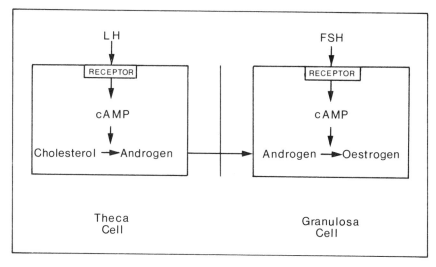

Fig. 2.21 The two cell, two gonadotrophin hypothesis for follicular oestrogen production. LH acts on theca cells to promote androgen production. The androgen is transported to the granulosa cells where it is converted to oestrogen under the action of FSH. This model may be oversimplified: granulosa cell aromatase, following induction by FSH, is probably able to respond directly to LH.

(Figure 2.21). There is also evidence to suggest that, following induction by FSH, granulosa cell aromatase will respond directly to stimulation by LH. The androgens, in addition to acting as oestrogen precursors, may stimulate the FSH-induced enzyme aromatase, which is involved in oestrogen biosynthesis. The oestrogen so produced stimulates granulosa cell proliferation. The follicle or follicles that eventually ovulate are probably those which, following gonadotrophin stimulation, most quickly acquire high levels of aromatase enzyme and subsequently develop LH receptors. Activin from granulosa cells may enhance the induction of aromatase by FSH, and may hence be involved in the conversion of gonadotrophin-independent follicles to gonadotrophin-dependent. As the follicle develops, the action of TGFα declines, but TGFβ becomes increasingly important, inhibiting granulosa cell proliferation and promoting gonadotrophin-induced steroidogenesis. In animals such as the cow, in which one or more follicles becomes dominant, gonadotrophin-stimulated production of inhibin by granulosa cells may act locally to promote insulin-like growth factor-I (IGF-I)-stimulated androgen synthesis by theca cells. This sustains oestrogen synthesis in granulosa cells of dominant follicles. In addition, oestrogen and inhibin act peripherally to block FSH secretion which is needed for continued support of non-dominant follicles.

At the culmination of preovulatory follicular development oestrogen passing into the peripheral circulation stimulates a preovulatory LH surge *via* a positive feedback mechanism (see page 19). This involves a switch of the hypothalamus from episodic output of GnRH to a huge surge in GnRH release; this in turn causes a surge in LH and FSH release from the anterior pituitary. The gonadotrophin surge initiates the completion of the oocyte meiotic maturation division (Figure 2.13), the rupture of the follicle [i.e. ovulation, by processes which probably involve steroidogenesis, increased levels of prostaglandin, plasminogen activator (which catalyses the conversion of plasminogen to plasmin), collagenase, and α-N peptide], and the process of luteinization, in which the cells remaining in the follicle after rupture differentiate, enlarge and become vascularized to form the corpus luteum (Figure 2.19).

Progesterone is the prominant product of the corpus luteum in all species. In some species significant quantities of other steroids (e.g. 17-hydroxyprogesterone and oestrogens) are also secreted, and the hormones required to maintain corpus luteal function vary. Luteinizing hormone is important in all species but maintenance of the non-pregnant corpus luteum varies, and may involve additional hormones

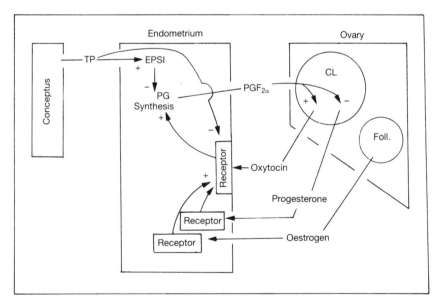

Fig. 2.22 Possible mechanism of corpus luteal regression in the absence of pregnancy and maintenance of corpus luteal function in early pregnancy, in ruminants. TP, trophoblastic protein; EPSI, endometrial prostaglandin synthetase inhibitor; CL, corpus luteum; Foll, follicle. (*After:* Thatcher, W.W., MacMillan, K.L., Hansen, P.J., Drost, M. (1989) *Theriogenology,* **31,** 151).

of a luteotrophic (corpus luteum maintaining) complex; these may include prolactin, FSH, and oestrogen. The maintenance of the corpus luteum depends on a balance between luteotrophic and luteolytic (corpus luteum regressing) mechanisms, the importance of which varies in different species. In the absence of pregnancy, corpus luteal function is not maintained due to either the absence of embryonic or placental luteotrophins or to the secretion of $PGF_{2\alpha}$ from the endometrium. This gains access to the ovary *via* diffusion from the uterine vein to the ovarian artery, with which it lies in close juxtaposition. The process is probably initiated by oestrogen from a developing follicle (Figure 2.22). The oestrogen acts to stimulate the production of oxytocin receptors on the endometrium. Oxytocin from the corpus luteum (or the posterior pituitary) then stimulates synthesis of $PGF_{2\alpha}$ by the uterus. This in turn decreases progesterone synthesis and stimulates additional oxytocin release. This sequence continues until the demise of the corpus luteum. The ability of the endometrium to synthesize and secrete $PGF_{2\alpha}$ in response to oestrogen and oxytocin is probably regulated by the duration of progesterone exposure, causing progesterone receptors to be lost by down-regulation and thus allowing the synthesis of oxytocin receptor. The regression of the corpus luteum in the non-pregnant animal leads to an initiation of preovulatory follicular maturation (for a detailed consideration of such oestrous/ menstrual cycles, see Chapter 3, pages 45–7).

2.6   EVENTS LEADING TO FERTILIZATION: FERTILIZATION

At mating, sperm are deposited at the top of the vagina or even (as in the pig) through the cervix (Figure 2.23). Usually, therefore, the sperm have to pass the cervix which, under conditions of progesterone dominance (luteal phase), will be blocked by a mucus plug. Under oestrogen dominance (during the periovulatory period) the mucus fibrils change their configuration and allow sperm passage. In some species the cervix, with its complex crypt structure, may act as a sperm reservoir. The exact mechanisms whereby sperm in the uterus find an unfertilized oocyte in the oviduct are unclear. Both sperm motility and uterine contractions may be involved in moving sperm up the female tract. What is probable is that of the millions of sperm deposited only a very small number of 'special' sperm may be permitted to the site of fertilization and that sperm are attracted by and have olfactory receptors for chemicals released by unfertilized oocytes.

During their passage through the female tract sperm undergo maturation: capacitation is the essential change which takes place in the outer membrane of the sperm prior to the events leading to fertilization (Figure 2.24). The initial interaction between the gametes involves

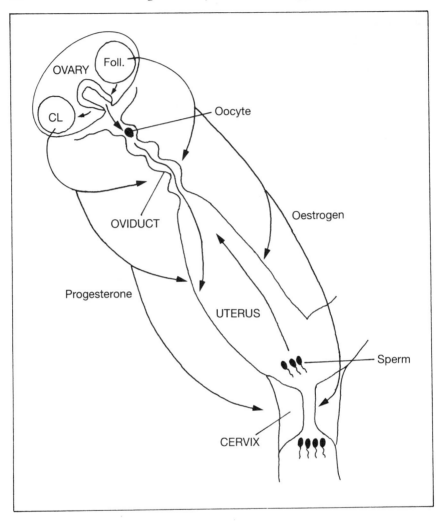

Fig. 2.23 Diagram of a female reproductive tract to illustrate sperm and embryo transport and the sequential influence of oestrogen and progesterone on the tract. CL, corpus luteum; Foll, preovulatory follicle.

binding of the outer membrane of the sperm to the outer surface of the zona pellucida (Figure 2.24). This triggers further events in the sperm head leading to the acrosome reaction, during which enzymes are released from the sperm head. Together with an increased motility of the sperm, these enzymes enable the sperm, now denuded of its outer membrane, outer acrosomal membrane, and acrosomal contents, to pass through the zona pellucida and into the perivitalline space.

The key events of fertilization are shown in Figure 2.25. After passing

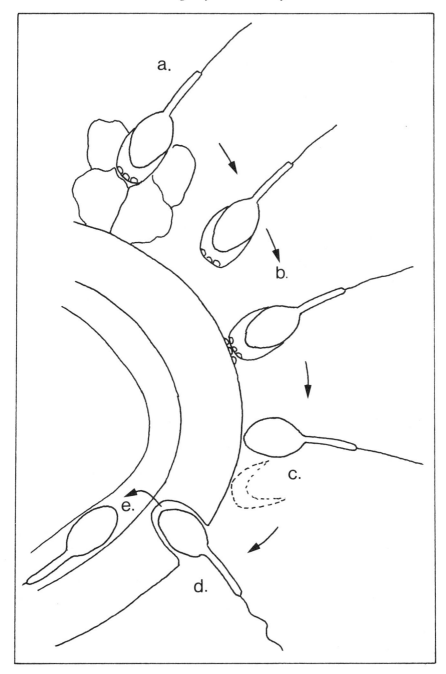

Fig. 2.24 The events leading up to fertilization. (a) Capacitation, (b) zona pellucida/sperm binding, (c) acrosome reaction, (d) sperm entering through the zona pellucida, (e) sperm in the perivitalline space.

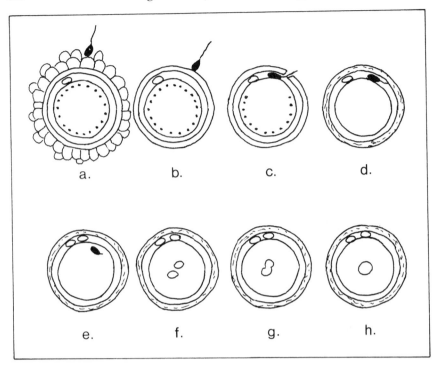

Fig. 2.25 Fertilization. (a) Oocyte after ovulation with first polar body; capacitated sperm pass through granulosa cells (see also Figure 2.24). (b) Sperm head binds to zona pellucida, enters the perivitalline space (see also Figure 2.24) and comes to lie against the oocyte triggering the cortical reaction (c) which in turn leads to the zona reaction (d). The sperm head sinks into the oocyte followed by the second meiotic division (e). Two pronuclei are visible (f) followed by the formation of a diploid nucleus (g) and activation for cell division (h). (*Note*: no abrupt removal of granulosa cells between (a) and (b) is implied; granulosa cells are gradually lost, and are not shown after (a) for simplicity).

into the perivitalline space the sperm head comes to lie against the oocyte (Figure 2.25c). This leads to a triggering of the cortical reaction, a release of the granules which lie within the cortex of the oocyte. This in turn leads to the zona reaction (Figure 2.25d). The latter is a change in the zona which prevents further sperm from entering into the perivitalline space to cause so-called polyspermy. Such a condition, in which the oocyte is 'fertilized' by more than one sperm, is not compatible with further development. The sperm head next sinks into the oocyte (Figure 2.25e), triggering the second meiotic division, which leads to the formation of a second polar body. Following a stage where two pronuclei are visible in the oocyte, the chromosomal material from mother and father come together, fertilization is complete, and the fertilized oocyte (zygote) is activated to undergo division (Figure 2.25h).

2.7 THE PREIMPLANTATION EMBRYO

After fertilization the dividing early embryo is transported to the uterus by ciliary action and by fluid from fallopian tube cells, under the control of increasing progesterone domination from the developing corpus luteum. A requirement of pregnancy in all eutherian mammals,

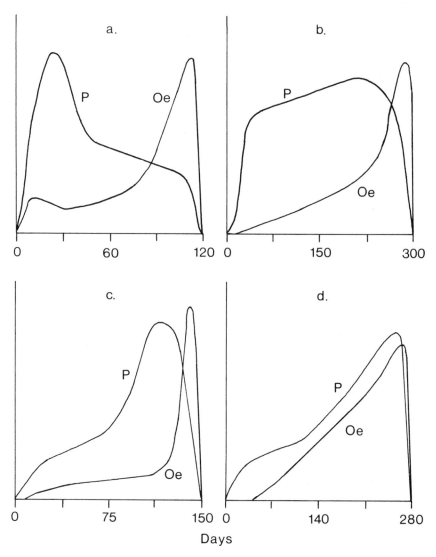

Fig. 2.26 Relative changes in progesterone and oestrogen during pregnancy. (a) Sow, (b) cow, (c) ewe, (d) man. (*Note*: absolute levels vary greatly between species).

at least initially, is that the life of the non-pregnant corpus luteum is extended (Figure 2.26). During pregnancy progesterone may be produced by either the corpus luteum or the placenta, or by both, depending on the species. Maintenance of the corpus luteum can involve a variety of signals from the developing conceptus, including (i) embryonic steroidogenesis, as in the pig where such oestrogen production directly or indirectly exerts an antiluteolytic effect by redirecting $PGF_{2\alpha}$ from the uterine vasculature towards the uterine lumen; (ii) antiluteolytic substances, such as the interferon-like tropho-blastic proteins of ruminants, which probably exert these effects by blocking mechanisms involved in $PGF_{2\alpha}$ synthesis (Figure 2.22); and (iii) embryonic or placental luteotrophins, such as human chorionic gonadotrophin (hCG).

The establishment and maintenance of pregnancy involves an array of embryo–maternal interactions, in addition to mechanisms concerned with maintenance of the corpus luteum. Production of interferon-like substances by the early conceptus occurs in species other than rumi-nants and may have a more universal role independent of luteal function. An early pregnancy factor (EPF) which may be immuno-suppressive has been reported very early after fertilization in a number of species, and the preimplantation embryos of some species produce platelet activating factor (PAF) which may stimulate EPF production and be involved in promoting implantation. On reaching the uterus the embryo may spend a more or less extended 'free living' phase within the uterus before the process of attachment begins, which is very different in different species. The uterine lining has been previously prepared for attachment by sequential stimulation by oestrogen prior to ovulation from developing follicles and progesterone from corpora lutea (Figure 2.23).

Progesterone alone is required for implantation in most species but oestrogen is also obligatory in some. The implantation process involves a complex local embryo–uterine interaction in which embryonic pro-teins, steroids, and other signals play a part. Eventually a placenta is established for respiratory exchange and nutrition of the developing fetus. The pregnancy is maintained by progesterone from the corpus luteum, the placenta or both, depending on the species. Details of implantation, pregnancy, and embryo/fetal development are outside the scope of this book.

2.8   PARTURITION

Parturition is a complex integrated series of events including; the softening and dilation of the cervix, possible pubic bone separation, the positioning of the fetus, the contraction of the uterine muscle, expulsion of the fetus, and the detachment and expulsion of the placenta.

Of particular interest is the initiation and timing of parturition. This results from a complex interplay of maternal, placental, and fetal endocrine interactions. These mechanisms vary between species, but in all probably involve the increased production of prostaglandin ($PGF_{2\alpha}$) for the stimulation of uterine muscle contractions.

There is evidence for a primary fetal initiation of parturition in some species which results in an increased ratio of oestrogen to progesterone in tissues. This in turn stimulates both increased prostaglandin production and an increased uterine muscle responsiveness to contractile agents ($PGF_{2\alpha}$, oxytocin). The possible universality of this general scheme (see Figure 2.27), which will be examined by reference to specific species, is complicated by the fact that the principal source of progesterone near term may be either the corpus luteum or the placenta.

The sheep is an example of a species dependent upon placental progesterone in the later stages of pregnancy. Hypothalamic stimulation of adrenocorticotrophin release from the fetal anterior pituitary causes an increased secretion of fetal adrenal cortisol. This activates the enzyme 17α-hydroxylase, thus reducing placental progesterone pro-

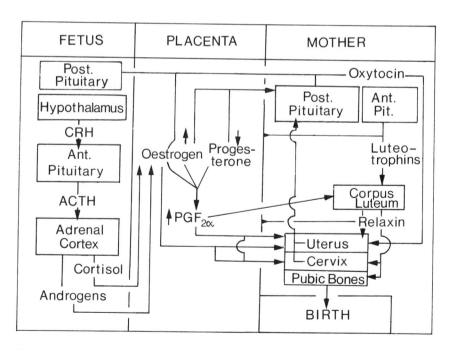

Fig. 2.27 Possible mechanisms for the hormonal initiation and control of parturition in different species. For detailed discussion, see text.

duction and enabling a rise in oestrogen biosynthesis (see page 16). This shift in oestrogen: progesterone ratio leads to an increased synthesis and release of $PGF_{2\alpha}$.

In primates the role of the fetus in the initiation of parturition is less clear: there is little evidence to suggest that the role of cortisol is the same as in the sheep. In man the role of the fetal adrenal in parturition may be to supply androgen precursors for oestrogen biosynthesis. There may not be a fall in progesterone before the onset of labour, and a change in peripheral oestrogen:progesterone ratio in the peripheral blood may not be required. There is evidence to suggest that oestrogen may inhibit progesterone formation to influence the local intrauterine oestrogen:progesterone ratio and promote $PGF_{2\alpha}$ production and uterine muscle responsiveness.

In species dependent on progesterone produced by the corpus luteum there is evidence that peripheral progesterone maintains uterine quiescence during pregnancy and that progesterone levels fall at term. In the goat activation of the fetal hypothalamo–pituitary–adrenal axis probably activates placental aromatase to convert fetal adrenal-derived androgens to oestrogens. This leads to increased $PGF_{2\alpha}$ production, which in turn causes corpus luteal regression. The degree to which $PGF_{2\alpha}$ may not be luteolytic, and to which the main luteal regression be caused by a withdrawal of luteotrophic support in some corpus luteum-dependent species is not clear. In the rat maternal influences, perhaps involving light and dark cycles and opioids, are important in controlling gestation length. In addition to their prime roles in uterine muscle and in luteolysis, prostaglandins are also involved in cervical softening prior to birth; oestrogen and relaxin have also been implicated. Relaxin is also involved in pubic symphysis relaxation in some species and, in the rat at least, may affect pregnancy length by inhibiting uterine muscle contractions both directly and indirectly by stimulating the opioid release which inhibits oxytocin release. The precise role of oxytocin in the initiation of parturition in different species is controversial. Oxytocin may stimulate the release of prostaglandins, and may stimulate uterine contractions directly. Mechanical stimulation of the uterus, the cervix, and the vagina, and a high oestrogen:progesterone ratio cause oxytocin release, suggesting that direct oxytocin effects may be secondary to those of $PGF_{2\alpha}$. Oestrogen has been shown, however, to increase the number of oxytocin receptors; hence increased action of oxytocin may precede rises in the circulating hormone. Oxytocin may also be produced by the fetus. A general model of the various mechanisms controlling the onset of parturition in different species is shown in Figure 2.27.

CHAPTER 3

Rhythms and limitations of reproduction

Infertility may be defined as a temporary inability or reduced capacity to reproduce. The permanence of such a condition is sterility. In the case of man infertility may be involuntary, as in other species, or voluntary by the use of contraception. Until recently male infertility has been studied less than female infertility, both in man and in other species. There has been a reluctance of men to accept responsibility for infertility as it is associated with reduced manliness. Andrology departments have also tended to be less well supported than gynaecology departments. In farm animals, while it is very important for all females of a herd or flock to be fertile, only a few males are required and infertile animals can be rejected. Finally sperm maturation from stem cells to ejaculated sperm takes weeks, and mechanisms controlling spermatogenesis are also concerned in libido maintenance. In man, therefore, the preovulatory follicle selection/ovulatory phase of women, which does not involve marked changes in libido, was an easier target for hormonal contraceptive development (see Chapter 7, page 139).

Conception occurs in only about 25% of cycles in fertile couples and couples are usually regarded as infertile when unprotected intercourse does not produce a pregnancy after 1–2 years. There are probably 60–80 million infertile couples worldwide: though clearly not health threatening, this can have an important social impact. Infertile couples may express disbelief, frustration, and anger. In studying human infertility problems it is important, where possible, to study couples from the onset of an investigation, as it may not be clear which partner is affected. About 30–50% of human infertility affects the woman, about one-third affects the man, and the remainder affect both partners. Ovulatory problems and tubal blockage are important causes of infertility in all women but there are large regional variations in their incidence. Tubal blockage probably accounts for about half of the infertility seen in African women compared to 20–30% in other women worldwide. Asian women have the lowest incidence of ovulatory problems and Latin American women the highest. Female infertility increases with age (Table 3.1): the proportion of presumably fertile married

Table 3.1 Number of married women who are in-fertile at different ages

Age (Years)	Infertile (%)
20–24	7.0
25–29	8.9
30–34	14.6
35–39	21.9
40–44	28.7

(*Data from*: three national US surveys *from* Menken, J., Trussell, J., Larsen, U. (1986), *Science*, **23**, 1389)

women, not using contraceptives, who fail to conceive in 1 year steadily increases from age 25 to age 44.↲

Failure of either the male or the female to reproduce is costly to the livestock farmer. Wastage in dairy herds, for example, each year may be around 20%, one-third of which is due to reproductive problems. In addition, a large number of bulls of proven genetic merit are culled due to impotence. In females, an important parameter for assessing fertility is 'non-return', that is the percentage of animals that do not 'return' to oestrus after either natural mating or artificial insemination and are, therefore, probably pregnant. Such animals are, however, not necess-arily pregnant and may be displaying an anoestrous or suboestrous condition or may have become pregnant and subsequently lost their embryo(s).

The presence of specific infertility conditions in wild species is dif-ficult to ascertain. Animals kept in artificial or confined conditions and social groupings often fail to breed. As with domestic animals, success-ful breeding of large wild mammals in artificial proximity to man (e.g. wild life parks) depends on a full understanding of their reproductive biology and social structure, and the development of good manage-ment practices. Small populations of wild mammals isolated by human and other pressures also suffer from reduced genetic diversity due to a high level of inbreeding, leading to the possible selection of infertile traits.

3.1 NATURAL SUPPRESSION OF FERTILITY

3.1.1 Variations in reproductive rhythms within the mammals

Mammals show great variation in the patterns of reproduction (parti-cularly in females). Interspecies variations are seen in age at puberty, oestrous/menstrual cycle length, gestation length, litter size, seasonality of breeding, delayed fertilization, delayed implantation, delayed intrauterine development, spontaneous and induced ovulation, and the presence of a spontaneous or induced luteal phase. Such differences reflect adaptation to the environment, male–female interactions, and pregnancy lengths 'selected' to enable mating to take place and young to be born at an optimal time of the year. To understand some of these variations it is helpful to consider variations on a 'typical' oestrous/menstrual cycle as influenced by environmental effects and male–female interactions in different species.

3.1.2 The oestrous/menstrual cycle

At puberty, female mammals commence a rhythmic cycle of ovarian function characterized by a recurring sequence of changes in the ovary, accessory organs, and in behaviour of an unmated animal. In most mammals this cycle is externally and visually characterized by a short period of sexual receptivity, associated with the time of ovulation. The period between successive oestrous periods is termed the oestrous cycle. The length of the cycle is largely governed by the length of corpus luteum activity. During this luteal phase, progesterone suppresses the secretion of FSH and LH required for the stimulation

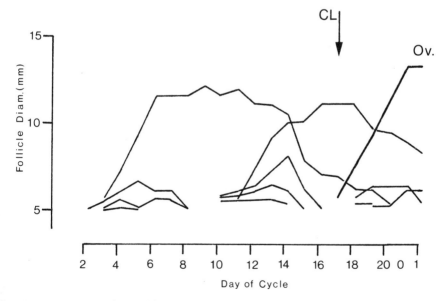

Fig. 3.1 Waves of antral follicular growth (greater than 5 mm diameter) during the oestrous cycle of the cow as seen by ultrasonography (CL, regression of corpus luteum; Ov., ovulation). (*Data from*: Jordan, J.E., Sirois, J., Turzillo, A.M., Lavoir, M. (1991) *J. Reprod. Fertil.*, **Suppl. 43**, 189).

of preovulatory follicular development. Only after corpus luteal regression will such gonadotrophin-stimulated preovulatory follicular development (and ovulation) occur. In some species, such as sheep and cattle, antral follicular development occurs outside this follicular phase, with two–three waves occurring during the luteal phase of the cycle (Figure 3.1), probably under basal gonadotrophin (FSH) stimulation. Final development of an ovulatory follicle does not occur, however, until after luteal regression.

A cycle containing such a period of preovulatory follicular development, a spontaneous ovulation with accompanying oestrus, followed by a spontaneous luteal phase, is shown in Figure 3.2a. Luteal regression is followed by a new period of preovulatory follicular development. An animal displaying a continuous series of such cycles is said to be polyoestrus (e.g. cow, sow). Other species have distinct breeding seasons that contain several such cycles if pregnancy does not occur; these animals are termed seasonally polyoestrus (e.g. ewes in temperate regions). The non-sexually active period is termed anoestrus. Some animals, such as the bitch, have one to three breeding seasons per year with only one oestrous cycle with each breeding season. She is termed seasonally monoestrus.

In old world primates, apes and women (Figure 3.2b) luteal regression is followed by a breakdown of the endometrium of the uterus and a loss of blood (menses). Externally in the resulting menstrual cycle the timing may be measured between bleeds rather than between oestrous episodes. The human is probably the only species which does not show oestrous behaviour at ovulation, though other primates may also show some receptivity throughout the cycle.

In some other species, female cyclic characteristics are influenced by the role of copulation in the induction of either ovulation or corpus luteal activity. Thus, in laboratory mice (Figure 3.2c) ovulation is spontaneous after preovulatory follicle development but the corpus luteum formed does not become fully functional unless mating occurs. Thus an unmated cycle is about 4–5 days long, whereas a mated cycle is about 12 days long (pseudopregnancy). In other species, sequential follicular development occurs, with the female being in a state of either continuous or intermittent heat, separated perhaps by periods of anoestrus. Ovulation only occurs following copulation. Thus such animals (e.g. rabbit, cat, camelids) are termed induced ovulators (Figure 3.2d). A variant on the induced ovulation system occurs in the field vole where coitus leads to ovulation but with an inactive luteal phase unless sufficient mating stimulus is given (Figure 3.2e). The distinction between induced and spontaneous ovulators may not be complete. Some measure of induced ovulation or critical ovulation timing may be induced by mating in apparently spontaneous ovulators.

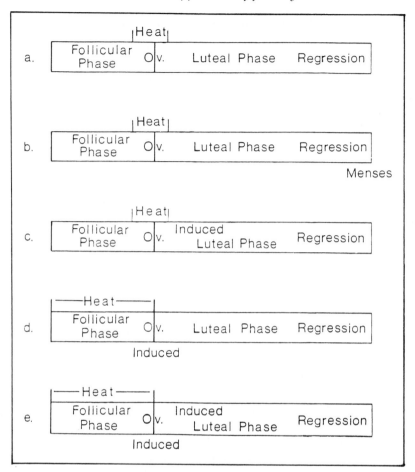

Fig. 3.2 Various types of non-pregnant ovarian cyclicity. (a) Oestrous cycle with a follicular phase, spontaneous ovulation and a spontaneous luteal phase. (b) Menstrual cycle with a follicular phase, spontaneous ovulation and a spontaneous luteal phase (note: the human shows no oestrous signs). (c) Oestrous cycle with a follicular phase, spontaneous ovulation and a male-induced luteal phase (pseudopregnancy). (d) Male-induced ovulation with subsequent luteal phase. (e) Male-induced ovulation with an active luteal phase only if sufficient mating stimulus is given (Ov., ovulation).

It should be noted that, although the variants of ovarian cyclicity discussed above are features of different mammals, in the wild such cyclicity will not normally occur, as the females will become pregnant. Repeated ovarian cyclicity may be regarded as a mechanism whereby a female achieves a second or subsequent opportunity for pregnancy if a first opportunity is not successful.

3.1.3   Periods of natural reproductive quiescence

During the lifetime of a mammal there are periods of natural reproductive quiescence: prepuberty, the interovulatory period of the oestrous/menstrual cycle, pregnancy, seasonal/environmental/nutritional anovulation and anoestrus, lactation and senescence. Suboptimal or intermittent infertility may also be experienced at the interface of some of these states: during the peripubertal period, at the beginning and end of a breeding season, or during the perimenopausal period in women.

The prepubertal/infantile period

The development of the gonads and their hypothalamo–pituitary control system occurs during the embryo–fetal stage, but full functional integrated development is suppressed until puberty, when the system becomes active by a number of mechanisms which are still not well understood. These may include changes in adrenal steroid secretion, changes in melatonin and opiate peptides and, above all, changes in the so-called 'gonadostat'. The hypothalamo–pituitary mechanism is very sensitive to negative feedback control by steroids in the young animal; as puberty approaches the sensitivity decreases, the activity of the hypothalamo–pituitary axis increases, and gonadal stimulation occurs. Nevertheless the gonads will respond to exogenous gonadotrophin prior to puberty.

The onset of puberty can be influenced by a number of factors, including the growth and metabolic status of the body, which in turn is influenced by nutrition. Undernutrition and poor body development delays puberty, for example in girls with anorexia nervosa; improved nutrition and faster attainment of body size advances puberty. Insulin-like growth factor I may be one of the metabolic signals involved in the initiation of puberty.

The attainment of puberty is gradual and difficult to define. In the male, the attainment of coitus with ejaculation of active sperm may be regarded as puberty; in women, ovulation is often not regular after the onset of puberty, which may be regarded as being complete only when ovulatory, regular menstrual cyclicity is established.

The interovulatory period of the oestrous/menstrual cycle

All spontaneously ovulating animals ovulate only at a single time in an oestrous/menstrual cycle which lasts a few days to a few weeks, depending on the species. Fertilization can, therefore, take place only during a limited period. The situation is different in species with induced ovulation (see page 46). There is some evidence in other species, including man, that some synchronization or induction of ovulation may occur to a limited degree.

Pregnancy and lactation

During pregnancy, oestrus and ovulation are normally suppressed due to the high concentrations of ovarian and/or placental steroids circulating at this time, blocking the hypothalamo–pituitary axis.

The antifertility effect of lactation has been recognized for many years. During lactation oestrus and ovulation are suppressed, though some species have an immediate postpartum oestrus, with early pregnancy and lactation occuring concurrently. In women, lactation is an effective inhibitor of ovulation. If bottle feeding is started immediately after birth menstrual cycles return much earlier than if an infant receives either a mixture of 'bottle' and breast feeding or breast feeding alone (Figure 3.3). Whether postpartum anovulation in different species is associated with lactation or suckling is not clear. In women, anovulation is associated with high blood prolactin concentrations in both postpartum and some non-lactating women (see page 54). In women who bottle feed their infant, prolactin levels return to normal much faster than in those who suckle their babies. In some species, however, the suckling reflex, associated with oxytocin release acting *via* an opioid inhibitory mechanism, seems to be the primary mechanism operating to inhibit oestrus and ovulation during lactation.

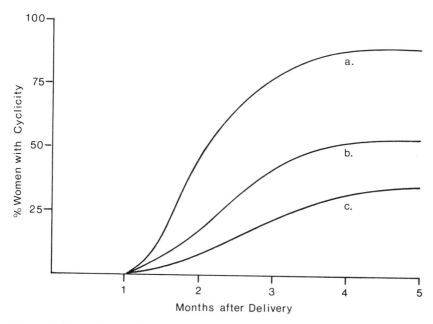

Fig. 3.3 The effect of lactation on the time of return of menstruation after birth in women (a) bottle feeding, (b) mixed feeding, (c) breast feeding. (*Data from*: Mastroianni, L., Coutifaris, C. (1990) *FIGO Manual of Human Reproduction Vol 1*, Carnforth, Parthenon, p. 84).

Seasonal and environmental effects

In many species of mammals gonadal activity in both males and females is continuous after puberty, apart from during pregnancy and lactation, until senescence. Other species display a marked pattern of reproduction with periods of activity being spaced by periods of inactivity. Species which are not truly seasonal (e.g. cattle and pigs, at least in temperate regions) may display periods of reproductive inactivity associated with seasonal climatic changes. At higher latitudes this onset of reproductive activity in true seasonal breeders is frequently associated with changing daylength. Such animals may be either 'shortday' breeders (e.g. sheep, goats) where decreasing daylength (autumn) is associated with the onset of reproductive activity or 'longday' breeders (e.g. horses, hamsters) where increasing daylength (spring) is associated with the onset of reproductive activity. These adaptations ensure that the young are born at a time to exploit a good food supply in the spring: autumn-mated ewes have a 150 day pregnancy and give birth to spring young, while spring-mated mares, with a 330 day pregnancy, give birth to spring foals. There is evidence that the pineal gland plays a key role in the perception and response to daylength changes, with the output of melatonin corresponding to the length of the night. Ewes probably have an endogenous circannual rhythm of reproduction which occurs without photoperiod changes. Reproductive transitions are generated by this endogenous rhythm, and the times of the changes are synchronized by the seasonal change in daylight, mediated by melatonin. Melatonin may act either on dopamine and opioid pathways or *via* the pars tuberalis of the pituitary gland to influence GnRH and gonadotrophin release.

Other climatic factors such as temperature, rainfall, and altitude may also be important controllers of reproductive function, both inside and outside the tropics. In the males of many species spermatogenesis requires a lower temperature than that of the abdomen, and elevated temperatures inhibit spermatogenesis. Such lower temperatures are often achieved by the situation of the testes outside the abdomen in a scrotum; in some species the testes may be maintained at a lower temperature within the abdomen by means of a complex counter-current blood flow system. The direct effect of climatic factors on the animal may be less important than their affect on food supply. Temperature and rainfall can dramatically affect the quantity and the quality of food available to unhoused, non-artificially fed herbivorous mammals.

Stress, whether environmentally or socially (e.g. overcrowding) imposed can affect reproduction. In general, stress-induced increases in adrenal function block reproductive activity, although in some circumstances modest stress may have a stimulatory effect (e.g. in pigs).

Prolonged exercise, associated particularly with runners undergoing intensive training, may also suppress gonadal activity.

Socially induced infertility is present in many species. High ranking or dominant individuals may suppress the reproduction of low ranking or subordinate individuals. Parents may also suppress the reproductive ability of maturing offspring. An extreme case is where a female accepts help in rearing offspring to maximize their survival chances (e.g. elephant). Such suppression may or may not involve pheromones. Certain social cues between individuals of the same species affecting reproduction certainly involve the transfer of pheromones. The addition of the ram to a flock of ewes in late summer hastens the onset of the breeding season (the 'ram effect'), putting a boar in an adjacent pen will hasten puberty in gilts (the 'boar effect'), and a mount-inducing pheromone has been found in the vaginal mucus of oestral heifers. Various toxic chemicals, including herbicides and pesticides, may disrupt reproduction. Such chemicals may persist in the environment or be concentrated in the food chain.

Nutritional effects

Optimal reproductive performance is achieved when the diet is balanced for energy, protein, vitamins, and minerals and meets the requirements for growth, maintenance, pregnancy, or lactation, as appropriate. Specific nutrients may also be important: lack of vitamin A, for example, affects spermatogenesis. As noted earlier (see page 48), there is a close relationship between nutrition, growth rate, and size at puberty, with undernutrition delaying puberty and improved nutrition advancing puberty. Restricted nutrition probably delays puberty *via* an inhibition of the pulsatile release of GnRH and consequently inhibition of gonadotrophin release. Decreased nutrition can also lead to adult anovulation and acyclicity. In women, amenorrhoea occurs when body weight falls to around 75% below ideal weight. Such weight loss may be associated with anorexia nervosa, a syndrome involving the adoption of various food phobias.

Improved nutrition in some animals can increase ovulation rate. This effect may be either chronic or acute, which is often referred to as 'flushing'. Higher ovulation rates reflect the induction of a greater number of large ovarian follicles in the ovaries of animals in good physical condition. Nutrition also plays an important role in the initiation of postpartum ovarian activity. Precisely how nutrition interacts with the reproductive control mechanism is not clear. However, the metabolic hormone, insulin, can increase ovulation rate; ovarian follicular granulosa cells have receptors for IGF-I; and growth hormone may have a synergistic role in both ovarian and testicular function. These observations suggest possible mechanisms for the mediation of stimulatory nutritional effects. The initiation of reproductive activity which follows increased feeding of growth-restricted lambs may be due to the increased availability of tyrosine. A reduction in growth and survival of embryos can occur under severe and extended periods of undernutrition. Paradoxically a high level of feeding in early pregnancy can also be detrimental, probably due to a decrease in progesterone caused by the stimulatory effect of high food intake on hepatic blood flow and progesterone metabolic clearance rate.

In addition to nutrition *per se*, other environmental factors may affect the ability of an animal to hunt or find food. Its reproduction may be affected by a nutritional deficiency even though the food supply is abundant.

Senescence

Reproductive function becomes impaired in old age, particularly in females, when the oocytes in the ovary are eventually expended. Perhaps only in the human does the natural life expectancy usually cover this phase of development. The menopause (strictly the stopping of menstruation) and reproductive senescence are more complex than a simple lack of oocytes. Prior to a lack of oocytes, women experience a period of suboptimal ovarian function, with irregularities of ovulation and cyclicity. Such irregularities involve changes in the hypothalamo–pituitary and feedback mechanisms controlling follicular development and ovulation.

3.2 SPECIFIC CAUSES OF INFERTILITY

In wild animals, domestic animals, and man, providing favourable conditions (e.g. good nutrition, environment, and social interactions) and reducing unfavourable conditions (e.g. antagonistic environment, stress, poor nutrition), leads to optimum reproduction. For farm animals this involves good management practice and stockmanship. Such environmental and nutritional factors, natural genetically determined reproductive rhythms, and individual congenital defects and diseases will therefore influence fertility.

For conception, pregnancy, and birth of live young to be possible an ordered sequence of events must occur (Chapter 2) involving sperm production, maturation and transport in the male; effective transfer of sperm to the female; further sperm maturation and transport in the female; synchronization of sperm disposition with ovulation in the female; fertilization, embryo maturation, and transport; endometrial development for implantation; maternal recognition of pregnancy and corpus luteal maintenance; embryo and fetal development; and successful parturition. Any of these key events or mechanisms controlling them can fail. In general the specific mechanisms involved have been studied in great detail in man, less so in domestic animals, and very little in wild species.

3.2.1 Failure of sperm to reach the oocyte

Male infertility

This may result in failure to produce and ejaculate sperm, premature ejaculation, or lowered libido with inability or unwillingness for intercourse. One in 25 men are subfertile; in 90% of these the ejaculate contains too few sperm (oligospermia), sperm of poor quality (low motility or sperm defects), or no sperm (azoospermia). Spermatogenic failure in man is only rarely caused by inadequate stimulation of the testes by gonadotrophins. This is seen, however, in patients with an extra X chromosome (Klinefelter's syndrome). A similar condition has been described in the ram and the bull. Cryptorchidism (failure of the testes to descend into the scrotum) results in inhibition of spermatogenesis by the relatively high abdominal temperature.

Spermatogenic failure has also been associated with mumps, with working with heavy metals, and with preparing cotton seeds containing gossypol (see Chapter 7, page 146). Men and animals with azoospermia may have obstruction of the vas deferens or epididymis caused by infection and inflammation, or a varicocele of the spermatic vein. Defects in testosterone production lead to low libido and impotency in man and domestic mammals. Unwillingness or inability to mate is probably the most frequent cause of infertility in domestic mammals. This may arise from factors other than lowered libido, including pain from genitalia or limbs during mounting, or an irregular or slippery floor surface.

Sperm—cervical mucus interaction; defective capacitation/fertilization, tubal blocks, adhesions

Sperm deposited in the female tract may fail to reach the fertilization site or fail to mature within the female tract. Hostile mucus within the female tract is responsible for about 5% of human infertility. Defective capacitation and fertilization are difficult to quantify, but they probably account for a number of cases of infertility of an unknown cause. A specific acrosomal abnormality has been demonstrated in bulls. A very significant cause of infertility in women (about one-third) is tubal blockage as a consequence of tubal infection or fibroma. The condition may be related to venereal disease, post-abortion infection or the use of an intrauterine device. It may be associated with chronic pelvic inflammatory disease (PID) or endometriosis. The latter is a syndrome in which the lining of the uterine endometrium grows at ectopic sites, which bleed at every menstrual bleed.

3.2.2 Ovulation failure and luteal phase defects

Failure of ovulation may be associated with primary ovarian failure, which may be partial or total. This is seen in premature menopause in women, or in those lacking one X chromosome (Turner's syndrome in women; related conditions have been described in the mare and the sow). This abnormal chromosome condition results in the development of 'streak' ovaries containing no oocytes: no folliculogenesis, oestrogen secretion, secondary sexual development, or ovulation is possible. Other ovulation or luteal phase disorders, at least in women, may be secondary to abnormality affecting the hypothalamo–pituitary feedback system, such as hyperprolactinaemia (high prolactin secretion which may involve a pituitary tumour) or disorders of thyroid or adrenal function. This also results in inadequate folliculogenesis.

Ovulation failure may occur in women and domestic animals due to the presence of follicular cysts. In cattle these may produce either excess oestrogens, resulting in a constant oestrus/nymphomanic condition, or excess androgens, causing virilization. Virilization in women and animals may also have a variety of other causes. In cattle the so-called Freemartin is a female calf which becomes virilized, often by androgens passing from a male twin due to anastomosis of the fetal placental blood systems. A hyperandrogenic syndrome with anovulation seen in women is probably caused by a spectrum of abnormalities; the syndrome includes patients with the classical Stein–Leventhal polycystic ovarian syndrome. A typical such patient is obese, hirsute, has polycystic ovaries, is anovulatory, and displays no menstrual cycles (amenorrhoea). Possible hormonal interactions involved in this condition are as follows. Elevated blood androgens, originating perhaps from the adrenal glands or ovaries, can be metabolized to oestrogens by the peripheral fat in an obese subject. An action of androgens on the liver may lower the production of sex hormone-binding globulin (SHBG). Hence, there will be an increased concentration of unbound biologically active oestrogens in the blood. Prolonged action of the latter on the hypothalamo–pituitary axis leads to low FSH secretion and intermittently high LH secretion from the anterior pituitary, leading to the development of cystic follicles producing high androgen levels from thecal tissue which is little aromatized *in situ* due to a lack of FSH action on granulosa cells. The excess androgens produced cause the hirsutism and continue the abnormal hormonal condition.

Luteal phase defects are associated with either too little progesterone production or with the development of a corpus luteum which regresses prematurely (short luteal phase). Such defects occur particularly at interfaces of reproductive inactivity and activity (e.g. between the breeding and non-breeding season, at puberty, or after parturition) and are associated with a lack of progesterone from a previous cycle. They can be associated with inadequate follicular development and, at least in women, are associated with hyperprolactinaemia and hypo-thyroidism. Ovulation can also occur without oestrus ('silent heat'). The first ovulation of a new season in ewes is characterized by such a condition, because normal oestrous behaviour is only shown when oestrogen action is preceded by progesterone. Such a sequence does not occur in the first cycle after anoestrus as no corpus luteum of a previous cycle is present at the first ovulation.

So-called luteal cysts can occur in farm animals, causing an extension of the luteal phase. These may be associated with uterine infections and the prevention of onset of luteolytic mechanisms from the uterus (see Chapter 2, page 35).

3.2.3   Implantation failure and embryo loss

The endocrinology of implantation is largely unknown in many species. The presence of an inadequate endometrium will prevent implantation. This may be due to the presence of a short luteal phase or to inad-equate maintenance of the corpus luteum of pregnancy because of an inadequate conceptus signalling mechanism (chorionic gonadotrophin or trophoblast protein) or, for example, follicular development (in ruminants) leading to an initiation of luteolytic mechanisms. Such factors may also be involved in embryo loss. Embryonic and fetal death, mainly involving early embryos, accounts for a substantial loss in both man and animal production. The exact extent of losses are difficult to assess (some women for example may never have known that they conceived) but up to 50% of early embryos may be lost. Some of the loss is due to failure of embryonic development due to a con-genital abnormality. Subsequent abortion may be due to disease, chro-mosomal, or immunological causes.

3.2.4   Parturition problems

Dystocia is defined as difficult birth. The problem may range from a slight delay or problem in the process to a complete inability to give birth or birth of a dead offspring. The placenta is usually shed after the expulsion of the fetus, but placental retention may occur.

CHAPTER 4

Reproductive manipulations: what to manipulate and what to use

4.1 OBJECTIVES

The specific manipulative method employed will depend on the specific objectives and species concerned.

In general procedures will be attempting to either: block reproductive function, usually for family planning in man and for certain specific reasons in domestic and wild species; or assist reproductive function. This may be required to induce 'normal' reproduction in man or animals with specific reproductive abnormalities, or to overcome natural reproductive limitations imposed, for example, by season or lactation, or to manipulate reproduction in order to increase productive efficiency, increase genetic potential, or conserve endangered species.

4.2 GENERAL LOCI OF ACTION

Figure 4.1 shows some of the key components of reproduction and its control systems in the male and female, and their interactions. Any point in these processes is a potential point of intervention for artificial reproductive control, whether stimulatory or inhibitory. The key to success is clear diagnosis or designation of the problem to be solved, followed by an attempt to find specific and reversible intervention methods. It is clear that detailed information on the reproductive processes allows the point of intervention to be more refined. Future approaches to both reproductive stimulation and inhibition will favour specific, perhaps local actions, rather than broad 'sledgehammer' approaches. Non-invasive, 'natural', perhaps pheromonal methods should be investigated further and advances in knowledge of autocrine and paracrine hormone actions, membrane biology, and immunology should be explored. No methods are likely to be applicable for all species

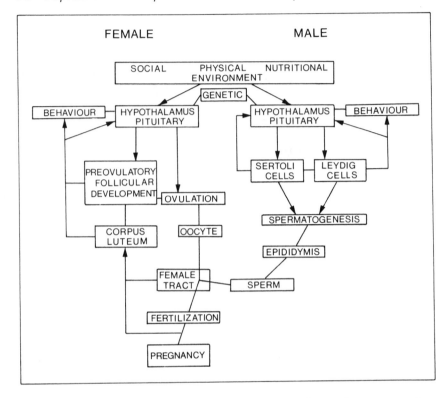

Fig. 4.1   Possible loci of action for various methods of reproductive control.

under all conditions. Hence, a wide array of possibilities should be investigated to supplement or replace existing procedures.

4.3   SPECIFIC LOCI OF ACTION

4.3.1   Manipulation of environmental influences on reproduction

The importance of environmental signals and influences on the control of reproduction has been stressed (see Chapter 3, pages 50–2). Factors such as light, nutrition, social cues, temperature, and stress can potentially be either exploited or minimized for the non-invasive, non-drug management of reproduction. Such methods may be more acceptable ethically and overcome the increasing worries of potential drug side-effects and the persistence of residues in the food chain. Some environmental factors are difficult to change. Changes in day length, for example, can undoubtedly be used to alter the breeding season of certain seasonally breeding temperate mammals such as sheep, goats and deer. Such animals need, however, to be housed in expensive, light-controlled accommodation. Other factors such as nutritional 'flushing' of ewes or exploitation of the 'ram' and 'boar' effects are much easier to apply. Good husbandry in domestic animals and the maintenance of habitat, social groupings, and age/sex balance appropriate for a wild species may be the most effective ways of maintaining reproductive efficiency. Particular regard needs to be paid to these features for large wild or endangered species that are being forced by man's intervention into progressively less natural conditions.

4.3.2 The use of known mediators of environmental effects

The influence of daylight changes on seasonal breeding phenomena are mediated by changes in melatonin (see Chapter 3, page 50). Changes in seasonal breeding may be achieved by the use of exogenous melatonin or its analogues. Various social cues, (e.g. the 'ram' and 'boar' effects) are transferred by means of pheromones. Once such pheromones are identified the individual chemicals may be used as appropriate stimulators or inhibitors of reproduction. Other such mediators of environmental effects on reproduction may be awaiting discovery.

4.3.3 Physical barrier and blockade methods

The most effective contraceptive method is abstinence (in the human) or separation of the sexes. Any physical or psychological barrier will also be potentially effective. These include interrupted coitus, vaginal douching or spermicides, condoms, diaphragms, caps, and blocking or cutting of the vas deferens or oviduct. The extreme physical methods are embryo or fetal removal and castration or ovariectomy.

4.3.4 Integrative brain–hypothalamo–pituitary mechanisms

It is becoming clear that a major common component in the control of reproduction is the GnRH pulse generator (see Chapter 2, page 23). Its specific nature is not yet fully understood. It may, however, be regarded as the sum of the integration of endogenous brain rhythms, environmental inputs, steroid and peptide feedback, neurotransmitters, opiate peptides, and other mechanisms leading to the stimulation or inhibition of GnRH synthesis and release, and/or the alteration of the sensitivity of pituitary gonadotrophin-producing cells to GnRH stimulation. Any or all of these mechanisms are potentially exploitable for reproductive manipulation under appropriate conditions. Reproduction may be blocked by the negative feedback of steroids and/or peptides, opioids, or the GnRH-induced down-regulation of pituitary gonadotrophin-producing cells. Reproduction may be stimulated by exploiting the positive feedback effect of oestrogen on gonadotrophin release, by the removal of negative feedback by active or passive immunization with steroids or peptides, by administration of an inactive receptor blocker such as an antioestrogen (e.g. clomiphene), an opiate peptide antagonist (e.g. naloxone), or neurotransmitter precursors capable of crossing the blood brain barrier (e.g. DOPA, a precursor of noradrenaline and dopamine). Gonadotrophin-releasing hormone and its analogues and exogenous gonadotrophins may also be used directly to stimulate gonadotrophin synthesis and secretion or gonadal activity, respectively.

4.3.5 Mechanisms controlling reproductive behaviour

In addition to manipulation of the mechanisms controlling the production of oocytes and sperm, the hormonal mechanisms involved in reproductive behaviour can also be controlled. In the case of the male the control of spermatogenesis and androgen production are directly linked (see Chapter 2, page 27). Such a relationship is also present in female mammals, where the display of oestrus, and the induction of ovulation involve oestrogen production. In the human, which is without oestrus, reproductive behaviour is influenced by adrenal androgens.

4.3.6 Gamete maturation, transport and fertilization

Hormones and other factors acting locally on the gonads in an autocrine or paracrine manner are currently under intense investigation. Such mechanisms may be exploitable for gonadal manipulation. Factors influencing sperm transport and sperm maturation in the epididymis (e.g. androgens and antiandrogens), the transport and maturation of sperm in the female tract (e.g. cervical mucus, female tract fluid components), and fertilization are all potentially manipulatable. Of extreme interest are possible manipulations of factors concerned with oocyte–sperm recognition and interaction prior to fertilization.

The transport and metabolism of both the unfertilized oocyte and the developing preimplantation embryo are influenced by oviductal and uterine mechanisms, in turn influenced by ovarian steroids and local signals, yet poorly understood, within the uterine lumen, which initiate implantation and other embryo–endometrial interactions. All such mechanisms may be potentially manipulatable.

4.3.7 Pregnancy

Factors involved in the regulation of pregnancy or in the demise of the corpus luteum in the absence of a developing conceptus have extensive potential for stimulatory and inhibitory intervention. Progestogens and antiprogestogens can be used to maintain or block pregnancy, while factors involved in parturition can also be exploited for abortion induction and the control of parturition.

4.4 HORMONAL AND RELATED CONTROL METHODS

Chemicals used to modify reproductive function may require application in very different ways to drugs that are used to affect disease or nutritional deficiency. They may have to be applied at different stages of a circadian cycle or at specific stages of the reproductive or life cycle for optimal effects. There are marked differences between different species both with respect to the timing and control mechanisms for different reproductive events (e.g. length of the oestrous/menstrual cycle, mechanisms controlling corpus luteal function). Different or even species-specific hormones may be involved. Oestradiol-17β, for example, may be the principal oestrogen produced by developing follicles of most species, but oestrone may be the main such steroid in some. Gonadotrophins from different species and even within species vary greatly in their half-lives in blood (see page 62): horse LH has a much longer half-life than sheep LH or cattle LH in the circulation.

4.4.1 Specific hormone and hormone-like preparations used

The hormones and related substances used for stimulation or inhibition of reproduction may be either:

(a) Naturally occurring compounds (see Chapter 2, pages 11–16) which have either been synthesized (e.g. steroids, GnRH), genetically engineered (e.g. recombinant FSH), or purified or semipurified from glands or fluids (e.g. pituitary gonadotrophins).
(b) Naturally occurring analogues (e.g. chorionic gonadotrophins, eCG and hCG, for use as mimics of pituitary gonadotrophins) or synthetic analogues (e.g. GnRH and prostaglandin analogues), or substituted compounds (e.g. steroid esters). Such compounds may be more active, have a longer biological half-life in blood or be orally active compared to the natural compounds.
(c) Stimulators or inhibitors of hormone synthesis, secretion, blood concentration or actions (e.g. specific antibodies and receptor blockers).

Naturally occurring hormones and analogues

Gonadotrophin-releasing hormone, the natural decapeptide (Chapter 2, page 13) of the hypothalamus, has been synthesized along with various analogues with alternative amino acids substituted into the molecule to produce potent GnRH agonists and antagonists. Potent agonists are obtained by substitution of the natural peptide at positions 6 and 10. Potent antagonists have been produced by substitution of up to seven of the 10 natural amino acids.

Melatonin (see Chapter 2, page 13) is the natural hormone of the pineal. Various analogues (e.g. naphthalene derivatives) have been synthesized.

The gonadotrophins FSH and LH are available as pituitary extracts from different species with varying degrees of purity and varying FSH/LH ratios. Genetically engineered products are also available. These potentially can be produced 'to order' with respect to individual hormone potency and half-life. Gonadotrophins are also available as extracts of human menopausal urine (human menopausal gonadotrophin, hMG), from the chorion of various species (e.g. the predominantly LH-like hormone from the urine of pregnant women; human chorionic gonadotrophin, hCG), and from the serum of pregnant mares (equine chorionic gonadotrophin, eCG, often called pregnant mares' serum gonadotrophin or PMSG). The latter has both FSH and LH-like activity. The half-lives of the latter two compounds in blood are higher than those of pituitary FSH and LH and, under certain circumstances, single rather than sequential injections can be used. This can, however, result in over-stimulation.

Pure synthetic steroids are available and various analogues and esters have been derived which are either orally active or have a longer bioactivity in the body. Oestrogens may be used as the natural compound (e.g. oestrone, oestradiol-17β, Figure 4.2a), as a derivative (e.g. ethinyl oestradiol, Figure 4.2b), as an ester (e.g. oestradiol benzoate, Figure 4.2c), or as a synthetic analogue (e.g. diethylstilboestrol, Figure 4.3b). Androgens may be used as the natural compound (e.g. testosterone, Figure 4.2d), as derivatives (e.g. methyltestosterone), or as esters (e.g. testosterone propionate, Figure 4.2e). Progestogens may be used as the natural compound (e.g. progesterone) or as derivatives of either 19-nortestosterone (Figure 4.2f; e.g. norethisterone, Figure 4.2g), or 17α-hydroxyprogesterone (Figure 4.2h; e.g. medroxyprogesterone acetate, Figure 4.2i).

The development of possible activin and inhibin analogues is awaited with interest. Prostaglandin $F_{2\alpha}$ (Chapter 2, page 17) has a very short half-life in the blood and several long-acting analogues have been devised (e.g. cloprostenol, dinoprost, luprostiol). The trophoblast proteins and hCG and eCG have been purified and it has been shown that recombinant interferon-α_1 1 mimics the effects of the trophoblast proteins from sheep and cattle.

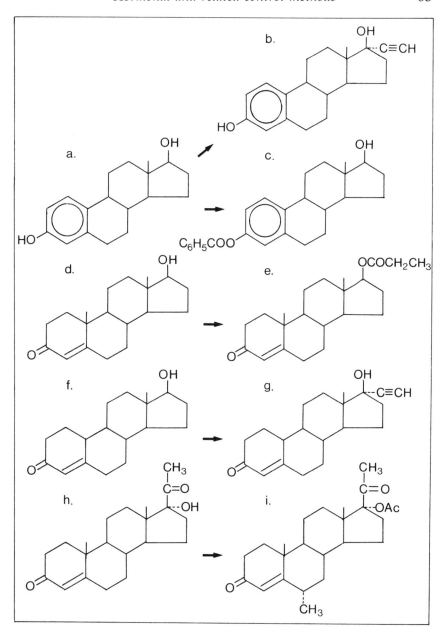

Fig. 4.2 Structures and derivations of certain synthetic steroids. (a) Oestradiol, (b) ethinyl oestradiol, (c) oestradiol benzoate, (d) testosterone, (e) testosterone propionate, (f) 19-nortestosterone, (g) norethisterone, (h) 17α-hydroxyprogesterone, (i) medroxyprogesterone acetate.

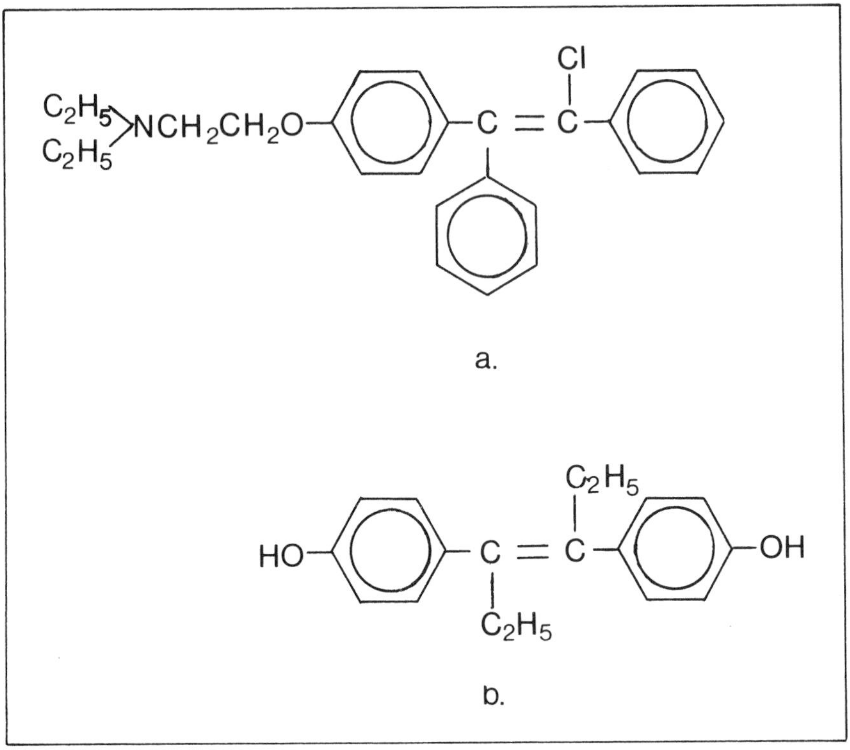

Fig. 4.3   Structure of (a) clomiphene and (b) diethylstilboestrol.

Stimulators and inhibitors

Several groups of potentially useful compounds are available for stimulating, inhibiting or for testing the integrity of reproductive function. These include:

(a) Drugs affecting neurotransmitters (and their metabolism and action) involved in GnRH synthesis and release; neurotransmitter precursors (e.g. DOPA, Figure 4.5b); and substances which mimic neurostransmitters (e.g. bromocriptine, a dopamine agonist).

(b) Opiate peptides and substances that block opiate peptide receptors (e.g. naloxone).

(c) 3β-Hydroxysteroid dehydrogenase inhibitors, hence progesterone synthesis blockers (e.g. epostane, Figure 4.4a).

(d) Anti-steroids, which act on specific receptors but have little biological action (antioestrogens, e.g. clomiphene, Figure 4.3a), antiandrogens (e.g. cyproterone acetate, Figure 4.5a) and antiprogestogens (e.g. mifepristone, Figure 4.4b).

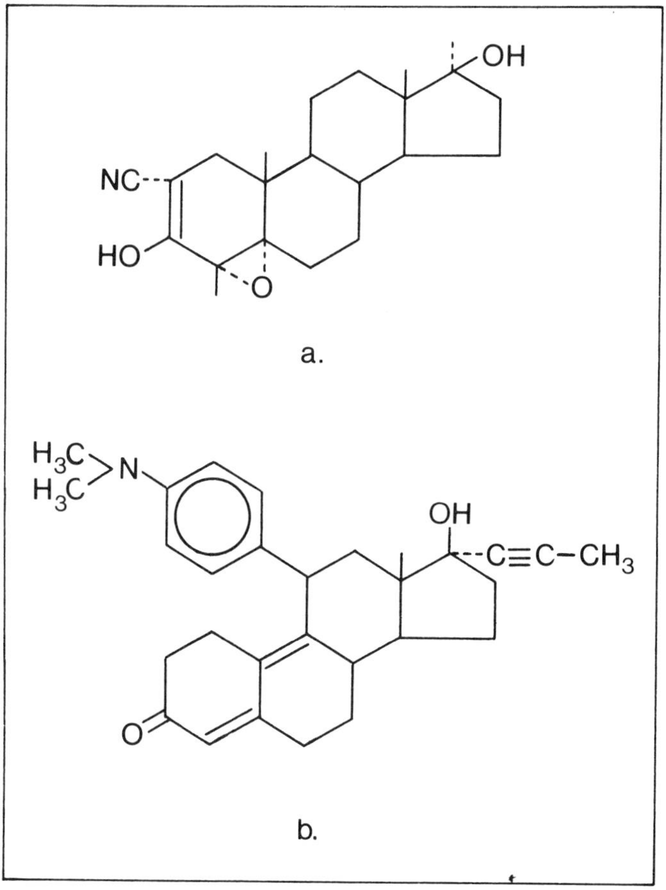

Fig. 4.4 Structure of (a) epostane and (b) mifepristone.

(e) Gossypol, a derivative from cotton seeds which blocks spermatogenesis.
(f) Active or passive immunization against hormones, including antibodies to gonadotrophins, steroids, and inhibins.

4.4.2 Route of administration, dosage, and half-life

Hormones and related substances may be administered by a variety of routes depending on the nature of the hormone and the type of effect desired (e.g. sudden increase, steady level etc.). A simple oral administration may be varied by using a nasal spray or a long-acting bolus. The latter may be in the form of a slow release formulation lodged in the rumen, for example. Intravenous, intramuscular, intradermal, in-

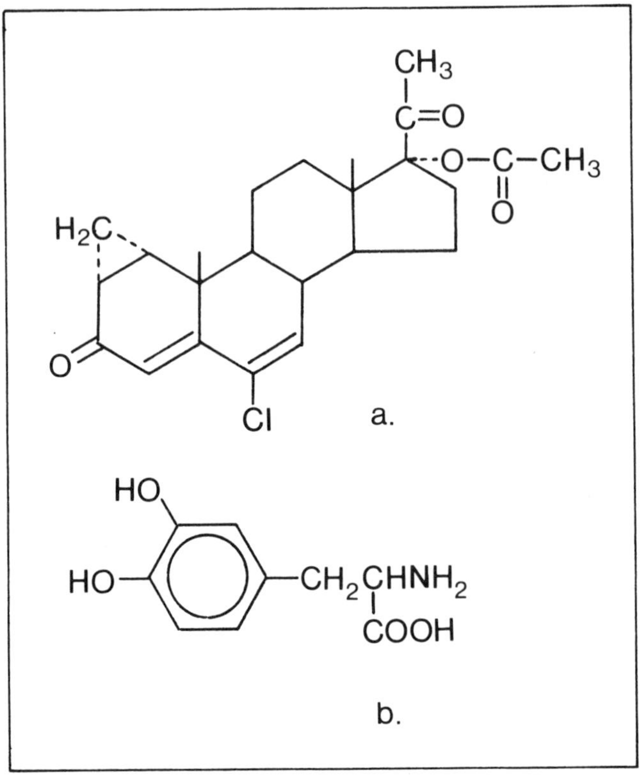

Fig. 4.5   Structure of (a) cyproterone acetate and (b) DOPA.

traperitoneal and subcutaneous injections may all be used in particular situations. Administration methods have been developed to avoid the need to give repeat injections or tablets. Such methods include topical application by means of a patch, the use of an implant, containing perhaps a long-acting or slow-release formulation, the use of a 'minipump' to achieve a continuous injection or a pulsatile input, or a vaginal ring or sponge to achieve a continuous effect. Devices such as the latter can be quickly removed, and they may thus be valuable for synchronization of cycles in animals (see Chapter 6, page 113).

The amount of hormone in the blood depends not only on the site of administration and the dose given but also its rate of elimination from the blood (see Chapter 2, page 16; and Figure 4.6). Such parameters vary with the hormone or analogue used and with the species, physiological state, age, and sex of the recipient. The metabolism of the hormone may result in the formation of another active or inactive compound which may be either retained within the body until further metabolism occurs (e.g. androgens peripherally metabolized to oestrogens in fat) or transported or broken down into a water-soluble

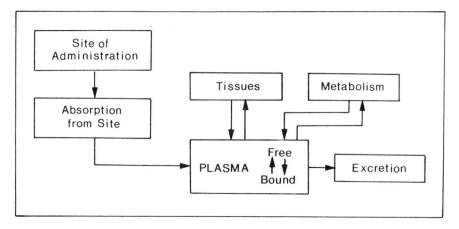

Fig. 4.6 The distribution of an administered hormone throughout the body.

product for excretion. The nature of the compound, its distribution within the body, whether it is retained in any tissues, and whether or not it is protein-bound in the plasma affect its metabolism and excretion.

4.4.3 Hormone and drug residues and risks

The use of hormones or drugs for disease control, growth or animal production, and reproductive manipulation in food animals is accompanied by the risk that residues of either the compound used, a metabolite, or a stimulated substance may be present in edible products derived from the animal. Such residues may be carcinogenic or teratogenic. There is also the risk that preparations derived from animal tissues may be sources of disease. Gonadotrophins extracted from pituitary gland may involve the risk of transfer of spongiform encephalopathy and related brain diseases. Creutzfeldt–Jakob disease may have been transferred in growth hormone preparations during the treatment of human dwarfism. The amount of a given hormone or drug that remains in the tissue after a given time will depend on a number of factors, including the route of administration, the type of hormone, the type of vehicle used, the metabolic clearance rate, the type of tissue studied (e.g. muscle, liver, milk), and the species.

Although hormones used for reproductive manipulation are usually not used in animals from which edible products are derived, withdrawal periods following the use of different preparations have been established to ensure that the hormone, drug, or metabolite will have been cleared from the animal's body before slaughter. Worries continue, however, and certain substances, such as growth-promoting steroids, are now banned within the EEC.

4.4.4   Problems in determining the efficacy of a compound

Problems associated with establishing the efficacy of a compound used to manipulate reproduction depend on the nature of the response expected and the situation and conditions under which it is administered. During research and developmental stages, a wide range of monitoring procedures are undertaken to determine specificity of action and side-effects. An established procedure is monitored by the overall efficiency of the result, plus a detailed analysis of any long-term problems or side-effects.

Specific, validated methods are needed to assess reproduction, not only to understand normal reproductive biology, but to diagnose abnormalities and to monitor the effects of manipulative procedures (see Chapter 5). It must be emphasized that there may be a wide variation of response to any procedure both between individuals and within individuals at different times. The response to a treatment may be expected to always have some sequential effects, such that an increased abnormality of response above or below the optimum may lead to an aberrant result.

CHAPTER 5

Methods of assessment of reproductive function

As well as understanding the features of normal and abnormal reproductive physiology of a species under investigation it is also important to be able to monitor the progress of events during procedures for stimulation or inhibition of reproduction. It is then possible to establish that the desired result has been achieved, and also to predict the successful (or unsuccessful) outcome of the procedure, and to monitor any deleterious side-effects. Even after a standard (e.g. fixed dose/fixed time) protocol has been established, shorter- or longer-term monitoring is important to confirm its continued efficacy. Modifications of a protocol may be needed to take account of individual variations of response.

Particular problems may arise in assessing infertile human patients. Conventionally, 1–2 years of failure to become pregnant in couples actively trying to conceive is regarded as infertility, but great variations may be present. The situation may be complicated by 'spontaneous' recovery from infertility: couples undergoing investigation or treatment, or even adopting a baby, may feel more settled and conceive their own child without help. A prolonged attempt to conceive by *in vitro* fertilization or other procedures may also increase stress and anxiety and lower the chance of success.

A wide variety of methods has been developed to assess normal and abnormal reproduction and the results of manipulation. Certain methods (e.g. the measurement of hormones in blood) may be applicable during developmental studies on man and domestic animals. Such methods may be less appropriate for use in human outpatients, under farm conditions, or in wild species. For the latter, problems of restraint and sampling have led to the development of methods such as urine sampling and behavioural studies.

For both human and other species it is usually important to obtain a detailed history and to undertake, if possible, a general examination which may include age, the reproductive history (e.g. age at puberty, previous pregnancies), and general lifestyle (human) or management (animal), including details of nutrition, environmental conditions, and general disease. For human infertility studies, in particular, it is also important to assess the male and female partners together. Failure of the woman to conceive may involve either or both partners or their interaction.

Methods for monitoring reproductive function must be reliable and practicable, and must relate as directly as possible to the function upon which information is sought. The frequency of observation or sampling and the interpretation of the data obtained is crucial. For example, if oestrous signs are observed as a parameter of synchrony of oestrous cycles (see Chapter 6, page 113) it does not necessarily imply that the oestrus is accompanied by ovulation. Similarly, the observation or palpation of a corpus luteum on the ovary or the presence of an increased plasma progesterone concentration confirms that a follicle has luteinized but does not necessarily confirm that ovulation has taken place, as luteinization can occur without ovulation.

5.1 OBSERVATIONS OF GONADAL AND ASSOCIATED FUNCTIONS

Gonadal function may be assessed in the male and female by direct or indirect observations of both gametogenic (ova and sperm) and hormonal parameters. In both sexes sexual maturity can be assessed by the development of secondary sex characters at puberty and by the development of such characters (e.g. antlers) with the onset of breeding activity in seasonal breeding species.

5.1.1   External signs in the female

Observations may be made (other than man) of signs of oestrous behaviour, including grunting, bellowing, and hyperactivity. Observations of such signs may be aided by the use of teasers (vasectomized males, aproned males, testosterone-treated females, oestrogen-treated females) to either elicit or detect oestrus. For sheep, ewes may be run with fertile or infertile rams with coloured raddle on the brisket so that at each mating the ewe will have a colour mark which can be changed (from yellow to blue to red) to allow sequential observations. In cattle, chinball markers, tailhead markers that rub off, or rump markers may be used. In sows, pressing of the back will evoke an immobility reflex at oestrus.

Such observations all monitor the hormonal status of the animal which induces oestrous behaviour. Since ovarian follicular oestradiol is the major component causing oestrus, oestradiol, and hence preovulatory follicular activity, is being monitored indirectly. Failure of menstruation to start (primary amenorrhoea), or temporary or permanent failure after it has started (secondary amenorrhoea) are probably the most noticeable sign of ovarian failure in women and other menstruating primates. The demonstration of amenorrhoea, however, gives little information as to its cause. Some indication can be gained by the use of a progestogen withdrawal test. In women with some follicular development and oestrogen production, the administration of progesterone can induce a uterine secretory endometrium and bleeding will occur when the progesterone is withdrawn. Other parameters for cycle detection, secondary to hormonal stimulation, are also employed. The increase in basal body temperature induced by the rise in progesterone in the second half of the menstrual cycle is frequently used in human and primates (Figure 5.1). Endometrial biopsies have been used to assess ovarian steroidal status. Changes in vaginal cytology, mucus discharge, electrical resistance of cervical mucus, mucus ferning pattern on drying, and mucus consistency (all of which alter under hormonal stimulation and hence indirectly assess oestrogen/progesterone status) have been used. Cervical mucus alters during the oestrous/ menstrual cycle, being thin, clear, and watery on oestrogen stimulation and thick and sticky during periods of progesterone dominance. Observations of cervical mucus can, therefore, be valuable in fixing the time of the cycle. Cervical mucus is 'receptive' to sperm for only a very short period of time around ovulation. In the majority of species, in which sperm are deposited in the vagina, the state of the cervical mucus influences sperm penetration, survival, and migration up the tract. Abnormalities of the cervix and its secretions are major causes of infertility in women.

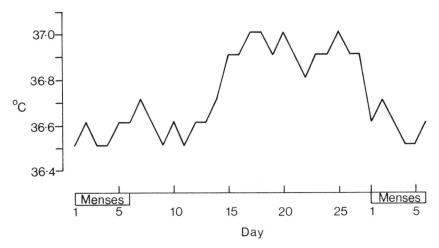

Fig. 5.1 The human menstrual cycle showing changes in basal body temperature.

The amount, viscosity, and cellularity of mucus may be recorded on arbitary scales. Viscosity may be quantified by observing the 'spinnbarkheit'. Mucus is placed on a slide, covered with a coverslip, and then drawn out between the slide and the coverslip. The length achieved before it breaks is measured in cm. The crystal formation of cervical mucus on drying is also characteristic of different stages: following its spreading on a slide, mucus is allowed to dry and the so-called ferning patterns examined.

The ovary can also be indirectly 'observed' by palpation (where possible, as in cattle) through the rectum and feeling for the presence of corpora lutea, cysts or large follicles.

5.1.2   External signs and observations in the male

Spermatogenesis

In the male the testes are usually outside the body in the scrotum; the size can easily be recorded by measuring the scrotum or by palpation. The majority of the testicular mass is seminiferous tubules, and testis size is highly correlated with sperm production in all species. Inactive spermatogenesis can therefore be diagnosed in infertile men and in immature or seasonally breeding animals. Firmness of the testis is also associated with sperm quality.

Libido

Libido in the male reflects the androgen status of the testis. In the female, ovariectomy or steroid replacement is acutely associated with the presence or absence of oestrous activity; in the male, however, loss and gain of libido after castration and androgen replacement may be more gradual. In the wild, complex rutting and fighting behaviour may be observed and in domestic animals libido may be tested by using teaser females.

5.1.3   Direct observation of the gonads

Direct observation of the gonads may be made, when appropriate, by laparotomy (surgical opening of the abdomen) or laparoscopy (observation through a small incision by a viewing laparoscope). Small samples may be taken at autopsy or by biopsy and prepared for microscopical examination. More detailed observations may be made of folliculogenesis, ovulation, and corpus luteum development by ultrasonography, which allows repeated non-invasive imaging of the

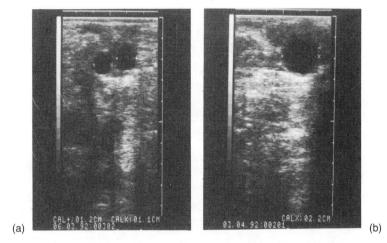

(a)

(b)

Fig. 5.2 Ultrasonic imaging of the ovary of a cow. (a) Ovary containing two large follicles. (b) Ovary containing one large follicle. (*Data from*: Tregaskes, L.D., Broadbent, P.J., Dolman, D.F., personal communication).

ovaries (Figure 5.2), reproductive tract, and developing fetus (see page 95). For ovulation assessment, for example, the increase in size of a large preovulatory follicle can be followed sequentially and the moment of ovulation noted by its collapse.

5.1.4 Observations of gametes

Assessment of oocytes and preimplantation embryos

Oocytes may be aspirated from antral follicles using laparoscopy and ultrasound, or following death, and oocytes or developing embryos may be flushed surgically or non-surgically (from uteri) at appropriate times after ovulation and insemination. Numbers of observed oocytes and embryos can be related to corpus luteal numbers. Embryos may be developed *in vitro* after fertilization *in vivo*. Oocyte maturity and integrity may be assessed by examination of the cumulus–corona complex of granulosa cells surrounding the oocyte and of the oocyte itself. Mature, preovulatory oocytes have an expanded luteinized cumulus and a 'radiant' corona (Figure 5.3). Intermediate oocytes have a less expanded cumulus; in immature oocytes the cumulus is unexpanded. Such a classification is somewhat subjective, and after gonadotrophin stimulation most oocytes may show some luteinization of

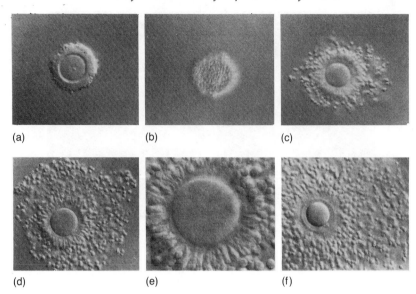

(a) (b) (c)

(d) (e) (f)

Fig. 5.3 Stages in the expansion of the cumulus oophorus, as seen by inter-ference contrast microscopy in mice. (a) Compact cumulus, (b) compact cumulus shown in tangential view, (c) expansion present at periphery, (d) expansion present in all but innermost cells, (e) higher magnification of (d) to show elongated corona radiata cells, (f) fully expanded cumulus (after ovulation). (*From*: Siracusa, G., De Felici, M., Salustri, A. (1990) in *Gamete Physiology* (eds R.H. Asch, J.P. Balmaceda, I. Johnston) Norwell, Serono Symposia, p. 137).

cumulus cells. The nucleus of the oocyte can be evaluated after gentle spreading of the cumulus to reveal the ooplasm and perivitalline space. The presence or absence of the first polar body and germinal vesicle can be determined to assess the nuclear maturation stage and possible abnormalities.

There is no ideal method for determining the developmental status and viability of the preimplantation embryo. Such embryos are often classified according to the integrity and appearance of the individual blastomeres. Thus, an excellent embryo will have blastomeres of equal size with no cytoplasmic fragments indicating blastomere disinte-gration. A good embryo will have only minor cytoplasmic fragments. A poor embryo will have unequal blastomeres and/or many fragments, and a bad embryo will have few or no intact blastomeres and much or complete fragmentation. The terms 'useable' or 'transferable' are frequently used, but these may vary depending on the purpose of the investigation and the species.

Embryo status and viability may also be determined by its ability to develop normally in culture, by the use of live/dead cell staining techniques, using fluorescent metabolic probes, and by microassays of culture medium, which measure embryo metabolism. The best viability test may be the non-invasive measurement of embryo respiration.

Assessment of semen

Spermatogenesis, the integrity of sperm transport, and the function of the accessory glands may be assessed by examining the ejaculate. The ultimate assessment of semen quality is clearly its ability to produce a pregnant female. Frequently, however, this will be an impractical, inappropriate, or expensive procedure. Seminal fluid characteristics may be affected by a range of social, seasonal and climatic changes, by stress, by breed, by age, and by the frequency of ejaculation. A number of *in vitro* tests have been developed with varying degrees of ability to predict fertility.

Classical tests for semen evaluation
These include measuring the ejaculate volume, the numbers of live sperm, the percentage of sperm with normal morphology and the percentage of motile sperm. Differences in ejaculate volume and sperm concentration for four species are shown in Table 5.1. Minimum normal values for a series of semen parameters in man are shown in Table 5.2.

Sperm numbers can be assessed either microscopically using special slides or counting chambers with marked grids, or with a spectro-photometer, an indirect method which relates sperm numbers to degree of light scattering (Figure 5.4). The ability of eosin or other dyes to penetrate damaged sperm membranes has been used extensively to distinguish between living and dead sperm: living sperm remain unstained, dead sperm are stained. Sperm with abnormalities of head or tail structure are found in all species, and suitably stained prep-arations of sperm may be examined to assess 'normal' morphology. Acrosome integrity can be assessed by specific staining. Motility (particularly rapid forward movement) is an important parameter of sperm quality. Such motility may be measured by counting motile and non-motile sperm, ideally under phase-contrast light microscopy, and

Table 5.1   Differences in ejaculate volume and sperm concentration

	Man	*Bull*	*Ram*	*Boar*
Volume of ejaculate (ml)	2–5	2–10	0.5–2	150–500
Sperm concentration (per ml $\times$ 10^6)	50–150	300–2000	2000–5000	25–350

Values given are ranges

Table 5.2 World Health Organisation criteria for normal semen analysis

Parameter	Minimum for normality
Volume (ml)	2.0
pH	7.2–7.8
Concentration sperm ($\times 10^6$/ml)	20
Total number of sperm ($\times 10^6$)	40
Motility 60 min after collection	
% with forward movement	50
(or) % with rapid linear progression	25
Normal morphology (%)	50
Viable (%)	50
Zinc (µmol/ejaculate)	2.4
Citric acid (mol/ejaculate)	52
Fructose (µmol/ejaculate)	13

Table 5.3 Motility scoring scale for sperm

Score	Assessment of movement
0	No forward movement
1	Weak, sluggish forward movement
2	Moderate, predominantly unidirectional
3	Good, unidirectional forward movement
4	Rapid, unidirectional forward movement

perhaps assessing them on an arbitrary scale (e.g. Table 5.3). More quantitative assessments may be obtained either turbidometrically or with photographic or computer-assisted microscopy.

The turbidometric method depends upon the ability of sperm to swim upwards from a concentrated cell suspension at the bottom of an optical cuvette into a layer of medium placed on top. Such movement results in a time-dependent increase in turbidity in the medium, which is proportional to the sperm motility. This can be measured by increased absorbance. The Makler counting chamber is a special chamber which enables the semen sample to be visualized over a field 10 µm in depth, allowing a sperm to be observed in a single perpendicular plane. Modifications of this chamber, with multiple exposure photography or videomicroscopy followed by either individual frame analysis or computer assisted analysis allow details of sperm movement kinetics to be analysed in addition to simple forward movement. Improving the precision of sperm motility measurement is not sufficient, however, and the conditions of the test must be standardized. The presence or absence of seminal plasma and the temperature are critical factors.

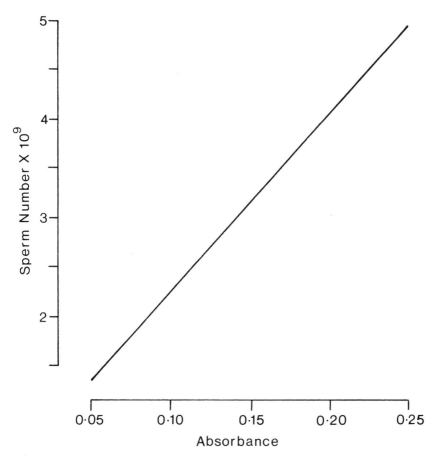

Fig. 5.4 Relationship between sperm numbers and optical density. (*Data from*: Ibraheem, M., personal communication).

Such tests for calculating the number of live forwardly motile sperm with normal morphology, although giving impartial assessments of sperm quality, are not directly predictive of fertilizing capacity. The increased use of artificial insemination, sperm freezing, *in vitro* fertiliz-ation and related procedures (see Chapter 6, page 117) have focused on the need to assess the integrity and fertilizing capacity of sperm directly, and to separate populations or even individual sperm of high quality.

Concentration techniques
A variety of techniques have been devised for the washing and concentration of motile sperm. The simplest are varients of the so-called swim-up migration methods and the use of filtration methods. In swim-up migration semen or washed sperm in the bottom of a tube

are overlaid with medium and incubated. The active sperm swim into the medium layer and can be collected. Sperm can be separated from acellular debris and leucocytes by filtration through glass beads or similar columns or by centrifugation through gradients of inert material such as Percoll.

Tests of sperm integrity

Tests have been devised to assess the outer cell membrane, acrosome, nucleus, midpiece, and contractile elements of sperm.

The hypo-osmotic swelling test measures the ability of sperm membranes to transport fluid. Sperm are incubated in hypo-osmotic media. If the membrane is functional water will enter, causing ballooning. This causes the contractile fibres of the tail to bend. The test is particularly useful to assess the possible effects of freezing. Chemical integrity of the acrosome can be assessed by the assay of acrosin, a protease probably involved in fertilization, whereas a sperm midpiece marker is creatine phosphokinase.

Sperm–mucus interactions

In many species sperm deposited in the vagina must penetrate the cervical mucus before entering the uterus. *In vivo* and *in vitro* tests of sperm–mucus interaction have been devised. The postcoital test is the only direct *in vivo* assessment of male–female interaction. A sample is collected after sperm deposition as near as possible to the time of ovulation and after previous abstinence from coitus for 2 days. The time of collection and source of the sample affects the interpretation. A vaginal sample will only show whether sperm were deposited. A cervical sample showing large numbers of forwardly motile sperm is positive, whereas small numbers of non-motile sperm will suggest a sperm and/or mucus abnormality. The interaction of sperm and mucus may also be investigated *in vitro* to test for both antisperm antibodies and mucus penetration by sperm. When sperm and preovulatory mucus are mixed, the presence of antibody is indicated by the presence of a high proportion of rapidly shaking motile sperm. Capillary tube columns of mucus to test for penetration of cervical mucus are available commercially. Movement of sperm up the capillary tube can be assessed.

Tests of sperm fertilizing capacity

The ability of sperm to express hyperactivation, a highly amplified and looping movement, is one of the best indirect methods of assessing sperm fertilizing capacity. Various substances have been shown to induce hyperactivity of sperm from different species (e.g. dibutyryl cyclic adenosine monophosphate, taurine), but no standardized tests are available.

Tests using either whole ova or zonae pellucidae from either the same or a heterologous species allow more direct assessment of fertilizing capacity. Tests using living eggs of the same species are clearly most indicative, but such eggs may not be available or appropriate to use. Zona-free hamster eggs have been extensively used to test for sperm penetrability by heterologous sperm. Homologous zona intact (living or non-living, fresh or stored) eggs can also be used. The ability of sperm to bind to and penetrate the zona pellucida does not depend on the viability of the oocyte. In fact the use of non-living eggs may be an advantage as the cortical and zona reactions causing the block of polyspermy (see Chapter 2, page 38) will not occur. Isolated hemizonae pellucidae can also be used to test for sperm binding capacity. The half zonae are prepared from spare or frozen oocytes or minced ovarian tissue, incubated with sperm, and the sperm binding is assessed microscopically.

Semen biochemistry

Only a small proportion (about 5%) of an ejaculate is sperm. The remainder comprises secretions from the accessory glands (e.g. the seminal vesicles, prostate and Cowper's gland), epididymis, and vas. The chemistry of semen varies greatly between species. Many of the components are derived exclusively or predominantly from one accessory gland. In man, for example, measurements of certain compounds will reflect activity of individual glands (fructose and prostaglandins, seminal vesicles; citric acid, prostate; carnitine, epididymis). In patients with azoospermia but normal spermatogenesis and endocrinology, suggestive of an obstruction, a lack of fructose in the ejaculate will indicate a block distal to the seminal vesicles or an absence of the vas, rather than a block in the efferent ducts of the testis.

5.1.5 Tests to assess the integrity of the female reproductive tract; hysterosalpingography, laparoscopy, hysteroscopy

In the female a full laparotomy will expose the reproductive tract. Hysterosalpingography is a technique which allows a radiographic display of the contour of the uterine cavity and of the patency or otherwise of the fallopian tubes. It is usually performed in women in the follicular phase of the cycle, 2–3 days after the end of menstruation. Laparoscopy allows visualization of the peritoneal cavity and, therefore, not only the observation of the ovary or reproductive tract, but also the possible diagnosis of abnormalities such as endometriosis and pelvic adhesions which may cause infertility. Hysteroscopy may be performed in patients with an abnormality demonstrated by hysterosal-pingography or in whom a direct observation of the uterine cavity is desired.

5.2   HORMONE DETERMINATIONS

Reproductive status can be assessed by the measurement of hormones produced both by the gonads and by the glands directly or indirectly affecting gonadal function (the pineal, hypothalamus, anterior pituitary, and uterus), as well as hormones of other glands (e.g. thyroid, adrenal) which may have a permissive effect on reproductive function. Which hormones to measure, and when and how will depend on a number of considerations, including the parameter to be assessed (Leydig cell function, follicle recruitment, corpus luteum function etc.); the nature of release (e.g. episodic) and clearance rate of the hormone; the nature of the hormone (e.g. existence of different forms, whether it is 'free' in blood, or a metabolite in urine); the nature of the test employed (e.g. basal measurement, inhibitory or stimulatory test); the nature of the sample (e.g. blood, urine, saliva, follicular fluid); and the methods available.

5.2.1   The parameter to be assessed

Methods are available for the measurement of the hormones produced by all components of the reproductive control system (e.g. pineal, hypothalamus, anterior pituitary, gonads).

5.2.2   The nature of hormone release and clearance

Many hormones are released episodically and each has a different rate of clearance from the blood. The frequency with which samples are taken for analysis has to take account of these: a single progesterone measurement may give a good indication of whether a corpus luteum has been formed, but frequent sampling is required to characterize the episodic output of gonadotrophins (Figure 5.5).

5.2.3   The nature of the hormone

Although testosterone may be the major active androgen in the male of many species, and oestradiol-17β is the active oestrogen secreted outside of pregnancy, other androgens are secreted by the testis and adrenals and other oestrogens may be secreted by the ovary. In some species and under certain physiological and pathological conditions

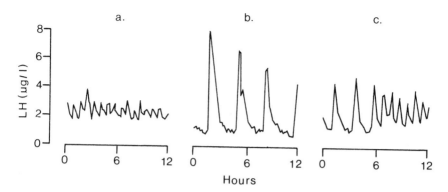

Fig. 5.5 Changing pattern of episodic secretion of luteinizing hormone at different phases of the oestrous cycle in the cow. (a) Day 3, just after ovulation/ early luteal phase. Many episodes of low amplitude are measurable. (b) Mid-luteal phase, day 10. There are few high amplitude pulses. (c) Day 19. As ovulation approaches increasing numbers of pulses are seen. (*Data taken from*: Rahe, C.H., Owens, R.E., Fleeger, J.L., Newton, H.J., Harms, P.G. (1980) *Endocrinology*, **107**, 500, 502).

these latter compounds may be the primary gonadal steroids. For example, oestrone rather than oestradiol may be of major importance in the non-pregnant llama. Gonadotrophins also exist in different forms under different physiological conditions. Differences in the carbohydrate component of these glycoprotein molecules alter their half-life and affect their apparent relative potencies in different biological and immunological assays.

Depending on the body fluid studied and the method employed, either the level of the 'free' non-protein-bound hormone, the total level of hormone, or the level of a metabolite may be measured.

5.2.4 The nature of the test employed

To assess endocrine function, using appropriate sampling procedures, either basal or hormonally manipulated concentrations of hormone may be measured. The interpretation of either a basal or elevated level may, however, be difficult. Endocrine dysfunction may result in either an under- or overproduction of a hormone or in an inappropriate response to a hormone. Thus a low level of gonadal steroid may imply gonadal failure, a low level of stimulation by gonadotrophins, or a low stimulation of the anterior pituitary by GnRH. A high level of gonadal steroid may imply a high degree of stimulation by gonadotrophins, or a tumour in the gonad directly causing the high concentration. The situation is illustrated in Figure 5.6, upper panel. An underfunctioning gland Y may either be lacking stimulation (Figure 5.6a) or may be

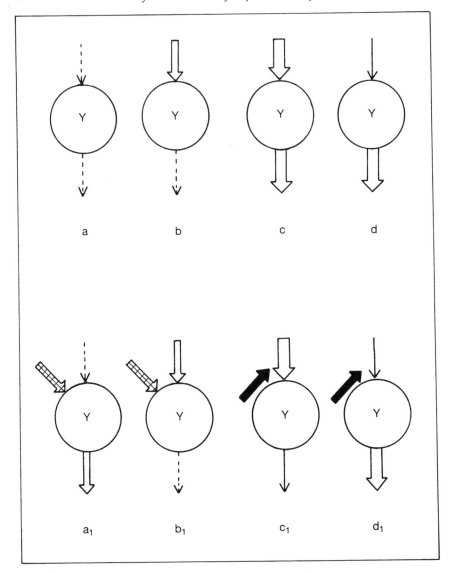

Fig. 5.6 Underfunctioning and overfunctioning endocrine glands and the diagnosis of the cause by the use of stimulation and inhibition tests. Upper panel: (a) underfunctioning caused by stimulation failure; (b) underfunctioning caused by failure to respond; (c) overfunctioning caused by overstimulation; (d) overfunctioning caused by autonomous overactivity. Lower panel: (a$_1$) stimulation test gives good response; (b$_1$) stimulation test gives no response; (c$_1$) inhibition test gives good response; (d$_1$) inhibition test gives no response.

adequately stimulated but be under-responding (Figure 5.6b). An overfunctioning gland may be over stimulated (Figure 5.6c) or be autonomously producing an excess of secretion (Figure 5.6d). The difference between these four situations cannot be determined by measuring the product of Y. A possible solution to the problem lies in the use of dynamic tests. These may take the form of stimulation or inhibition tests (Figure 5.6, lower panel). To distinguish between (a) and (b) an exogenous stimulator of gland Y may be applied. If the gland is lacking in endogenous stimulation but is functionally intact to respond then a good response may be expected (Figure 5.6a$_1$); if the gland is failing to respond to stimulation no response will be expected to exogenous stimulation (Figure 5.6b$_1$). Such a test using GnRH could attempt to distinguish between hypothalamic and pituitary failure of gonadotrophin release. The results of such a test are given in Figure 5.7. Men with severe hypogonadotrophic hypogonadism due to presumed hypothalamic failure may have a subnormal response to GnRH stimulation but may respond normally to such stimulation following continued infusion of GnRH. Such a protocol may allow a distinction to be made between hypothalamic and pituitary hypogonadism. Another such test is one in which the ability of the hypothalamo–pituitary axis to respond to positive oestrogen feedback and produce an LH surge may be tested by the administration of oestradiol followed by measuring the LH response. A parameter that cannot easily be tested directly (hypothalamic function) can therefore be indirectly tested. To distinguish between (c) and (d) (Figure 5.6) an exogenous inhibitor of gland Y may be applied. If the gland Y is overactive because of overstimulation then a block of this overstimulation (Figure 5.6c$_1$) will lower the activity of Y, whereas if the overactivity of Y is due to autonomous activity of Y then block of the stimulatory pathway to Y will not lower the activity of Y (Figure 5.6d$_1$). A well known example of such a test is the so-called dexamethasone test used to distinguish between patients with abnormal cortisol production due to an adrenal tumour or ectopic adrenocorticotrophin (ACTH) stimulation, or due to excessive stimulation from excess pituitary ACTH (Figure 5.8). Dexamethasone (a synthetic adrenocorticosteroid) given orally will block excess ACTH output from the anterior pituitary (*via* negative feedback) and hence cortisol output will be lowered (Figure 5.8a); if the excess cortisol is from an adrenal adrenocorticoid-producing tumour or stimulated by ectopic ACTH (Figure 5.8b and c), cortisol levels will not be lowered.

An additional way to overcome problems associated with interpreting basal data for the measurement of a hormone is to monitor sequential hormone profiles, such as LH over the period of an expected

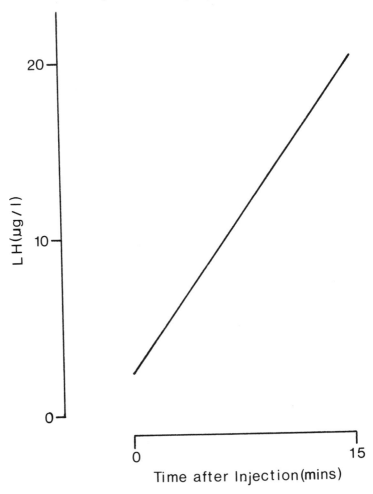

Fig. 5.7 GnRH stimulation test in the cow. The data show the mean LH response of four cows given 20 µg GnRH intravenously at time O. (*Data from:* Dobson, H., Alam, M.G.S. (1987) *J. Endocrinol.*, **113**, 169).

preovulatory surge (Figure 5.9), or daily changes in gonadotrophins and steroids to characterize a primate menstrual cycle or other mammalian oestrous cycle (Figure 5.10). The timing and frequency of sampling needs to be carefully controlled to take account of cyclical hormonal variation or episodic output.

The problem of interpretation of basal hormonal data can also be overcome by measurement of more than one hormone from a linked system. Thus if the hypothalamo–pituitary–Leydig cell system is considered (Figure 5.11a) both pituitary LH and Leydig cell testosterone can be measured. If low testosterone accompanies high LH

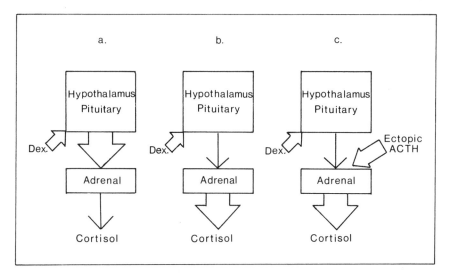

Fig. 5.8 The dexamethasone test. Dexamethasone will block excessive cortisol output caused by excessive stimulation from the pituitary (a). If excessive cortisol output is caused by an adrenal tumour, dexamethasone will not block its production (b). If excessive cortisol output is caused by an ectopic ACTH-producing tumour, dexamethasone will not block its production (c).

(Figure 5.11b) primary Leydig cell failure could be suspected, suggesting that without any presence of testosterone negative feedback, LH secretion will be high. If low testosterone accompanies low LH (Figure 5.11c) low testosterone levels may be due to an inadequate stimulation by LH, due to hypothalamo–pituitary failure. As can be seen from Figure 5.11d and e, if high testosterone accompanies high LH the inference is that excessive stimulation of the Leydig cells cannot be lowered by the negative feedback action of testosterone, suggesting a primary overproduction of LH, whereas if high testosterone accompanies low LH the inference is that feedback is operating normally and that the Leydig cells are not being excessively stimulated. Hence the conclusion would be that such Leydig cells are autonomously producing an excess of testosterone. In female patients with Turner's syndrome or postmenopausal women, in whom there is primary ovarian failure, ovarian steroid concentration is low but gonadotrophin concentrations are high, whereas patients with ovarian failure due to pituitary or hypothalamic failure have low levels of both ovarian steroids and gonadotrophins.

5.2.5 The nature of the sample for hormone assay

Prior to 1960 hormone assays to assess reproductive function were relatively insensitive and required large volumes of blood. For hu-

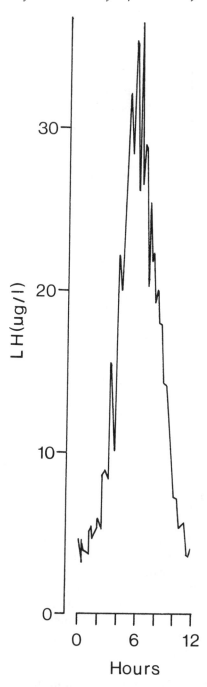

Fig. 5.9 Changes in plasma LH during the preovulatory surge phase in a cow. (*Data from*: Rahe, C.H., Owens, R.E., Fleeger, J.L. *et al.* (1980) *Endocrinol.*, **107**, 501).

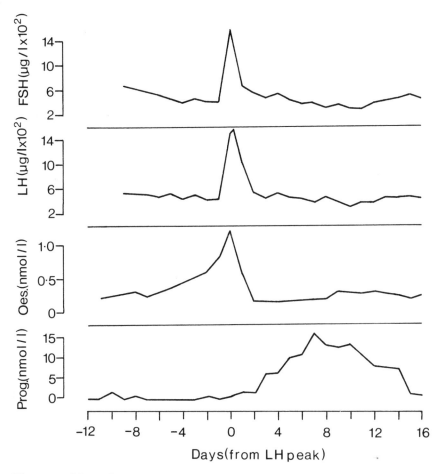

Fig. 5.10 Mean changes in FSH, LH, oestradiol and progesterone during the menstrual cycle of the rhesus monkey. Values are taken plus or minus the LH peak (LH and FSH standard – LER-M-907D). (*Data redrawn from*: Spies, H.G., Chappel, S.C. (1984) *In* Marshall's Physiology of Reproduction, 3rd Edition, Volume 1, p. 668. Ed. G.E. Lamming. Churchill Livingstone, Edinburgh).

man diagnostic purposes, practical methods were developed which measured levels of hormones or their metabolites in urine. The advent of radioimmunoassay (see below) and other related methods allowed assays to be developed with sufficient sensitivity to measure hormones in small volumes of plasma or other body fluids. Such plasma methods became the methods of choice and revitalized the understanding of reproductive endocrinology. More recently there has been an increased interest in the use of other body fluids and products, including saliva, urine, milk, and faeces for monitoring the hormonal status of human,

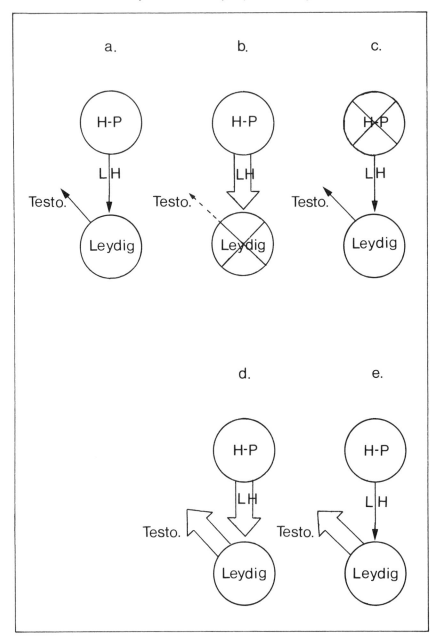

Fig. 5.11 Measurement of two components of a linked system (e.g. pituitary LH stimulating Leydig cell testosterone (a)). Low testosterone with high LH implies primary Leydig cell failure (b); low testosterone with low LH suggests hypothalamo-pituitary failure (c); high testosterone with high LH suggests primary overproduction of LH (d); high testosterone with low LH suggests that the Leydig cells are autonomously producing excessive testosterone (e).

farm and wild animals. Such methods do not involve invasive blood collection and plasma separation, and samples can be collected in the field or in a patient's home. Urine samples can be collected, for example, from rhinoceros or elephant 'scent marking' sites, and saliva samples for assessment of corpus luteal function in the assessment of human cycles can be taken daily by patients on 'filter paper' planchettes and kept in the ice box of their refrigerators for batch return to the clinic.

Data derived from different body fluids and products require careful interpretation. Hormones measured in milk, saliva, urine, or faeces may be metabolized, conjugated to sulphate or glucuronide, and perhaps represent the 'free' non-protein bound fraction of plasma. An elevated level of a hormone in urine may also bear a different time relationship to other physiological parameters than does the same hormone measured in plasma. In addition, a 24-h urine collection represents an integrated excretion value from perhaps a sequence of episodic secretions into the blood over the whole period of study.

5.2.6 Methods available for hormone assay

Before 1960 few routine assays for hormone concentrations in biological fluids were available. Biological assays performed on large samples of plasma, urine, or urine extracts were possible to a limited degree (e.g. urinary gonadotrophins to diagnose human hypogonadism and human pregnancy tests measuring hCG). Steroid hormones or their metabolites could be measured by physicochemical methods. The advent of radioimmunoassays and related methods allowed sensitive assays for all reproductive hormones to be developed for use in plasma and in other biological fluids and products.

Radioimmunoassay and related methods

The basis of a hormone radioimmunoassay is the competitive inhibition of the binding of a radioactively labelled hormone to an antibody by the unlabelled hormone. The basis of the reaction is shown in Figure 5.12. The essential components of such a system for carrying out an assay are:

(a) either a hormone of high purity for labelling (usually with ^{125}I) or a hormone preparation containing ^{3}H labels.
(b) A specific antibody, prepared either as a polyclonal antibody by the immunization of a test animal or a monoclonal antibody produced by preparing a clone of specific antibody producing cells *in vitro*.
(c) A method for separation of the antibody-bound hormone from the free hormone in order to separate antibody-bound radioactivity from unbound radioactivity.

$$\begin{array}{c} H^* \\ H \end{array} + AB \rightleftharpoons HAB + H^*AB$$

Fig. 5.12 The principles of radioimmunoassay. A radioimmunoassay is based on the competitive inhibition of the binding of a radioactively labelled hormone (H*) to a specific antibody (AB) by the unlabelled hormone (H).

Briefly, an assay is carried out by setting up a series of standards over the working range of the method and a series of unknown samples. A fixed amount of labelled hormone and antibody is added to each tube, including a series of control tubes, containing only labelled hormone and antibody and only labelled hormone. After appropriate incubation, antibody-bound hormone and free hormone are separated. A standard curve is plotted. An example of such a curve is shown in Figure 5.13a, in which the percentage of bound radioactivity (H*AB) is plotted against the logarithm of the hormone concentration of a standard. An unknown sample can be read off against the standard curve as shown. In practice, the standard curve data are transformed into a linear plot (Figure 5.13b) and calculation of concentrations of unknown samples is undertaken using appropriate computer software.

Non-radioactive labels which have been developed for use in assays rely on enzymes (in which the response is a coloured product) or fluorescent or phosphorescent endpoints. A complex array of assay systems have been developed including the so-called ELISA (enzyme-linked immunosorbent assay) method. Kits have been developed for use within and outside specialized laboratories, in the home, on the farm, and in the field. The potential for further advances is great. Miniaturized biosensors are being developed which may enable the continuous monitoring of hormonal concentrations by use of a small implant. The results from such devices can be recorded by the use of miniaturized radiotransmitters. Such monitoring devices could potentially be linked to hormone or drug administering devices so that hormone monitoring and manipulation could be linked.

Biological assays

Although immunoassays can give sensitive and precise measurements of hormones in body fluids, the results are dependent upon the cross-

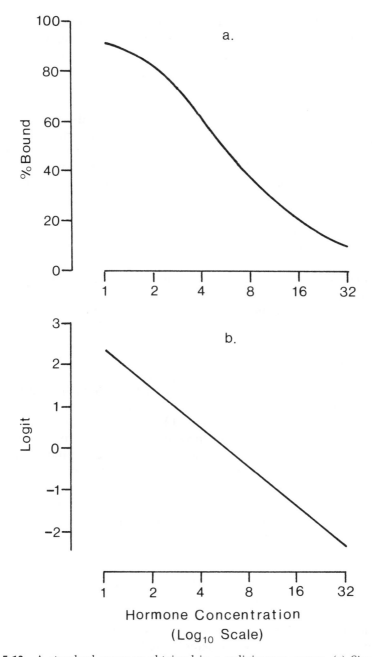

Fig. 5.13 A standard curve as obtained in a radioimmunoassay. (a) Sigmoid curve obtained by plotting the percentage of radioactive tracer bound to antibody (ABH*) against the $\log_{10}$ hormone concentration. (b) Linear plot obtained by transforming the radioactive tracer bound to antibody to the logit.

reactivity of the hormone with an antibody. Such an immunoactivity will not necessarily correlate with the biological activity of the hormone present, which can only be checked by the use of an appropriate biological assay.

In vivo bioassays

Hormone potencies of gland extracts and body fluids were originally determined by *in vivo* bioassays. These involve the measurement of changes of an appropriate endpoint of action of a hormone after its injection into test animals. Thus oestrogens can be measured by their effect on the uterine weight of immature mice (the mouse uterus test), and LH can be measured by its ability to stimulate testicular Leydig cells to produce testosterone, which can in turn be monitored by increases in ventral prostate weight in immature rats with their pituitary glands removed (the rat ventral prostate assay).

Such methods tend to be insensitive and cumbersome, but they do measure the total action of a hormone within the body of the test animal, including absorption from the injection site, metabolism within the blood, action at the target cell receptor, and the induction of the cellular response.

In vitro bioassays

Methods measuring the cellular response   Methods have been developed for hormone bioassay *in vitro* using tissues or cells, either in incubation medium or in culture. For example, LH may be assayed by measuring the testosterone production of incubated testes tissue or Leydig cells, and FSH may be assayed by measuring oestrogen production from cultured granulosa cells. Such methods are frequently more sensitive and more economical than *in vivo* methods and give an estimate of biological activity, receptor action, and subsequent cell response. They do not, however, take into account metabolism of the hormone in the blood of a test animal.

Receptor assays   Specific tissue receptors or plasma binding proteins may be used to assay hormones in systems similar to immunoassays, in which the antibody is replaced by a specific biological binding reagent. The specificity of the method is determined by the inherent specificity of the binding protein. Under certain conditions the results of such assays more clearly resemble the biological than the immuno-logical result. The use of a plasma binding protein is, however, essen-tially unrelated to the true biological action of the hormone and, even using a target cell receptor, only the initial binding of the hormone is examined, rather than its ability to initiate a full cellular response.

Thus, an oestrogen will bind to an oestrogen receptor within a target cell and initiate a response, whereas an oestrogen antagonist will bind to the receptor but initiate little or no cell response. Similarly an oestrogen agonist will bind to the receptor and initiate a cellular response.

Assay standards and validation

The importance of using an appropriate standard in hormone assays cannot be overemphasized. Where possible, different doses of the authentic substance (e.g. steroid) should be used for comparison. During studies of a 'new' hormone or where a chemically pure substance is not available (e.g. gonadotrophins), working standards, if possible ones which are made widely available, should be used. Such preparations are available for comparison with laboratory standards from the National Institute of Health, USA, the National Institute for Biological Standards and Control, UK, and the associated World Health Organization International Laboratory for Biological Standards. In some cases international standards (e.g. hCG, eCG) or international reference preparations (e.g. hMG) have been prepared. In the case of the gonadotrophins species-specific differences exist and heterologous standards must be used with caution.

All assays require careful validation and quality control. Methods must be checked for specificity (lack of interference by other hormones or non-hormonal substances), sensitivity (measurement of the hormone present in the specimen or extract), accuracy (measurement of the true level of hormone), precision (acceptable error), and parallelism (dilution curves of unknown samples are parallel to the dilution curves of the standard, implying that the 'same' substance is being measured).

5.3 DETECTION OF OVULATION

Much of the previous discussion in this chapter has included methods for the indirect or direct detection of ovulation. Methods for detecting ovulation and its timing are very important for natural and artificial hormonal inhibition of reproduction (see Chapter 7), for many methods of stimulating and synchronizing reproduction in females and for techniques involving artificial insemination (see Chapter 6). Such methods have, however, proved elusive, and as discussed earlier, are frequently indirect. Ultrasound can be used, but it is clearly limited by cost and cannot be used under many circumstances. Other methods often used include monitoring of 'mittelschmerz' (the ovulatory pain experienced by some women), detection of oestrus, changes in vaginal mucus and cytology, changes in basal body temperature, detection of the progesterone fall and of the preovulatory oestrogen and LH rises, or the postovulatory rise in progesterone. All are either indirect, retrospective, or unreliable. A reliable practical package of parameters for the detection of ovulation is needed for all species.

5.4 MISCELLANEOUS TESTS

Non-reproductive parameters may be involved in studies of reproductive function. Hence a study of reproductive abnormality may involve the study of adrenal and thyroid function, non-reproductive function of the pituitary, including growth hormone and prolactin secretion, radiographs of the pituitary fossa (in man, thinning of the floor of the sella turcica indicates a pituitary tumour), and parameters of general metabolic function.

5.5 ASSESSMENT OF PREGNANCY

It is important to be able to diagnose pregnancy as soon as possible, both in man and in other mammals. In farm animals, for example, the savings and practicalities of early remating and the removal of non-pregnant or barren stock may be very great.

5.5.1 Assessment of fertilization and implantation

Although early knowledge of fertilization and implantation is important, the detection of these events remains elusive. Successful fertilization *in vivo* is not detectable, though there have been hopes for the detection of early pregnancy factor (EPF, see Chapter 2, page 40) in several species. Failure to produce an endometrium which is receptive to the blastocyst and which will support implantation, either in normal reproduction or during embryo replacement or artificial placement (see Chapter 6, page 119), may represent a major cause of pregnancy failure. In the latter situations synchronization between the stage of development of the embryo and the endometrium may be at fault. In man at least, morphological assessment of the endometrium by means of a biopsy is possible. More practical and useful will be the demonstration of specific uterine substances that are either involved in implantation or are an endpoint of hormonal action of implanatation, which could be used as diagnostic tools. Such approaches are showing promise.

5.5.2 Pregnancy diagnosis

Pregnancy can be diagnosed by a number of means depending on the circumstances.

Observation of non-return

A missed period in women and higher primates or the non-return to oestrus in other mammals is good evidence of pregnancy. Embryo loss may, however, take place after this time.

Observation of pregnancy

Changes in behaviour and physiology (e.g. morning sickness in women), and changes in body weight and shape are clearly useful parameters for some purposes but may be too imprecise or too late for others. Early 'observations' of the conceptus may be made by palpation in some species and by ultrasonography (which has essentially replaced radiography), which is now in use for a wide variety of species. Various types of external or rectal probes are used, depending on the species.

Hormone assays

Assays for specific substances involved in pregnancy recognition (e.g. hCG) have been used for the detection of human pregnancy for half a century. The progressive development of assays for other pregnancy-specific substances will enable early pregnancy to be diagnosed. Progesterone assays have been extensively used in many species: in all eutherian mammals early pregnancy is maintained by an extension of luteal function (see Chapter 2, page 40). If a sample is taken within the period of potential return to oestrus (Figure 5.14), then a high concentration of progesterone can be diagnostic of pregnancy. If the timing of mating is not known accurately then (particularly if only a single sample is used) such a sample may coincide with increased corpus luteal function associated with a reovulation and a false-positive result will occur. Later in pregnancy other hormones give an indication of ovarian, placental or fetal function. In man, for example, placental lactogen is produced by the placenta and secreted in increasing amounts parallel to the increasing weight of the placenta (Figure 5.15). Oestriol production in the human reflects the ability of the fetal liver to hydroxylate the steroid molecule at position 16 (see Chapter 2, Figure 2.6, page 16). Hence normal oestriol production suggests normal fetal liver function and, indirectly, normal fetal well-being. The fetus may also be viewed by ultrasound or by fetoscopy.

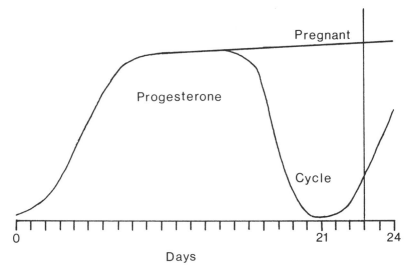

Fig. 5.14 Early pregnancy test in cattle measuring progesterone (*After*: Ministry of Agriculture, Fisheries and Food (1984) Dairy Herd Fertility p. 68).

5.6 FETAL ABNORMALITIES

Early embryo loss is common in many species: genetically and environmentally induced congenital abnormalities are an important cause. The majority of such abnormalities are usually lethal. In man only 5% of liveborn infants have significant or serious congenital abnormalities.

If abortion of such conceptuses is to be offered to the human mother early diagnosis of the abnormality is desirable, so that an abortion can be carried out as early as possible. Certain human abnormalities can be diagnosed by amniocentesis (the collection of fluid and fetal cells from the amniotic sac) at around 16 weeks gestation and by chorion villus sampling (Figure 5.16) at around 8–12 weeks. Amniotic fluid increases in α-fetoprotein, for example, indicate spina bifida, and three copies of chromosome 21 are present in infants with Down's syndrome. Specific genetic probes are making diagnosis of an increased array of diseases possible. It may be, however, that if suspected in a family such diagnosis could best be performed on preimplantation embryos after *in vitro* fertilization (IVF) procedures (see Chapter 6, page 117), thus avoiding the need for abortion of an implanted conceptus. These latter techniques may have roles in other species particularly with respect to the sexing of embryos.

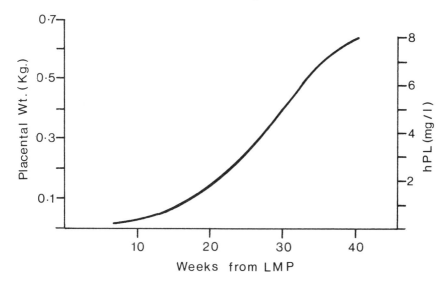

Fig. 5.15 Mean changes in placental weight and placental lactogen during pregnancy in the human. Note that the changes in the two parameters essentially fit an identical curve (LMP, last missed period). (*Data adapted from*: Hytten, F.E., Leitch, I. (1971) *The Physiology of Human Pregnancy* Oxford, Blackwells, p. 292; Boyd, J.D., Hamilton, W.J. (1970) *The Human Placenta* Cambridge, Heffer; Yen, S.S.C., Lein, A. (1984) in *Marshall's Physiology of Reproduction*, Vol. 1. 3rd Edn. (ed G.E. Lamming). Edinburgh, Churchill Livingstone, p. 742).

5.7 THE SELECTION OF SEX

The ability to preselect the sex of an offspring could greatly increase genetic improvement and the efficiency of livestock production. In man such selection of embryos during an IVF procedure could allow, in particular cases, the transfer of only embryos without a sex-linked abnormality.

5.7.1 Preselection of sperm

The only established difference between X and Y chromosome-bearing sperm is the quantity of DNA in the sex chromosome. In fact viable X and Y sperm populations have been separated according to their DNA contents in a flow cytometric/cell sorter. Widespread application of the technique is presently limited by the cost of the apparatus, the number of sperm that can be processed, and the increased embryo mortality which is related to the use of a fluorochrome to stain the sperm.

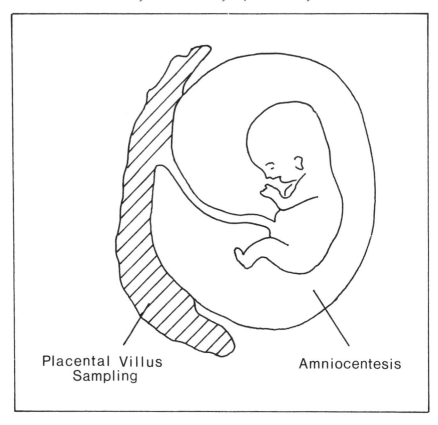

Fig. 5.16 Sampling of amniotic fluid and chorionic villi for the diagnosis of human congenital abnormalities. (*From*: Open University (1974) *Physiology of Cells and Organisms*, Unit 10, p. 16).

5.7.2   Selection of embryos

Various methods have been derived for the sexing of embryos. Cells from the embryos can be karyotyped, to establish the presence of an XX or XY chromosomal complement. The most promising methods involve the use of DNA probes specific to the Y chromosome, and the polymerase chain reaction, which provide accuracy, speed, and reliability.

5.8   METHODS OF ASSESSING EFFECTIVENESS, RISKS AND SIDE-EFFECTS

For the assessment of any contraceptive method, effectiveness, safety, reversibility, acceptability and freedom from side-effects need to be ascertained.

5.8.1 Contraceptive effectiveness

The effectiveness of any method is a quantitative measure of the ability to prevent conception. It is useful to compare different contraceptive methods for effectiveness in several ways:

(a) The theoretical effectiveness will equate very closely to the lowest observed failure rate (the failure rate observed when the method is used correctly under ideal conditions).
(b) The use effectiveness is a measure of the failure rate taken from a large group of users and will include a number of errors and 'human' failures.
(c) Cost-effectiveness is a measure of relative failure rate in financial terms.

These effectiveness criteria are measured in terms of percentage failure rate.

5.8.2 Contraceptive and conception assistance safety

Although the safety of conception assistance methods for use in man is very important, there is usually only a small risk from the treatment (see Chapter 6, page 129). Long-term use of contraceptives may present greater risks. It is important that the risks and benefits should be carefully assessed and that the risks should be equated with other chronic risks to which people are exposed, such as smoking or driving (see Chapter 7, page 149). Any new drug is subject to a series of trials. Toxicology tests are followed by a small ethically approved trial of human volunteers, followed, if appropriate, by larger trials to assess efficiency and side-effects. After acceptance of a drug for use, assessment continues to check its performance and any long-term complications that may occur. The overall risk of a method can be compared to other methods and to the risk of maternal mortality and morbidity associated with an unwanted pregnancy. In evaluating the acceptability of a method its continued use needs to be assessed in different cultural and religious backgrounds.

CHAPTER 6

Methods for reproductive induction, stimulation or assistance

Manipulation to induce, stimulate, or assist reproduction may be used in man, domestic and endangered/wild mammals for a variety of purposes. These include:

(a) *The induction of 'normal' reproduction*

Normal active reproductive function may be induced in males and females during periods of either abnormal infertility or natural reproductive quiescence, including prepuberty and seasonal and lactational anovulation. The ability to remove the period of seasonal anoestrus could greatly facilitate the production of marketable lamb throughout the year and reduce the non-breeding interval after lactation in cattle. This would enable the production of one calf per year to be more easily attained.

(b) *An increase in number of gametes (and offspring) to optimal numbers*

The number of offspring that a female can carry, the optimum litter size, is limited by the 'carrying capacity' of the uterus and the ability of the mother to sustain an increased fetal load. The number of offspring that can be cost-effectively carried may be higher, however, than the number that are usually carried. Thus, though twins are uncommon in domestic cattle, twins can be readily carried under good management conditions. Although a non-litter species cannot be turned into a litter species, an increase in the number of offspring may be possible and appropriate. A policy to achieve an optimal fetal load for the conditions may be the desired aim. In a temperate lowland environment, all ewes may optimally carry twins, whereas in poor highland conditions single lambs may be the norm.

(c) *An increase in number of gametes to supraoptimal numbers*

Such procedures may be carried out to obtain oocytes for *in vitro* fertilization (see page 117) or embryos for potential transfer to other individuals, either directly or after freezing (see page 126). They may also be used in connection with alleviation of infertility, increasing reproductive efficiency, or increasing genetic improvement.

(d) *The altered timing of normal reproduction in order to synchronize events*
When embryo transfer is being undertaken (see page 119) it is very important that the development of the reproductive tract of the recipient female is synchronized with the development stage of the embryo being transferred. Artificial insemination also requires ovulation to occur in the recipient at the time of insemination. In animals, it may be convenient for batches of recipients to be synchronized for transfer or insemination at the same time. In the human cycle such timing can be used for manipulative convenience.

(e) *Overcoming of inbreeding*
Isolated wild, zoo, or domestic mammal populations may become very inbred with the concomitant loss of hybrid vigour. Such problems may be overcome by maintaining gene pools, in the form of frozen sperm or embryos, from which genetic diversity may be reintroduced to a population.

(f) *The enhancement of embryo survival*
In all species there is a high incidence of embryo loss, particularly in early pregnancy. That which is not due to genetic embryonic deficiencies may be preventable.

(g) *The control of parturition*
In order to manage both normal and abnormal conditions of the mother and developing fetus it may be convenient or necessary to be able to induce early or timed births.

(h) *Hormone replacement therapy without fertility*
In some circumstances such as congenital hypogonadism (e.g. Turner's syndrome, see Chapter 3, page 54), premature or normal menopause, or for the production of 'teaser' male or female stock, it may be useful to give steroid reproductive hormone replacement, not to induce fertility, but to induce the development of normal secondary sex characters, prevent menopausal symptoms or to induce normal reproductive behaviour, respectively.

Methods for inducing, stimulating or assisting reproduction fall mainly into two groups:

(a) *The manipulation of hormonal control mechanisms*
This may be achieved using natural hormones or their analogues, drugs, immunization or other means to alter hormone function; altering genetic mechanisms controlling reproduction; or using environmental means including nutrition, photoperiod, social/pheromonal or other management procedures. In animals optimal reproductive function can best be maintained and much potential infertility be counteracted by good nutrition and management. In man overall reproductive health depends on good living conditions,

including food, responsible individual reproductive behaviour, and good basic and reproductive health care.
(b) *The manipulation of gametes*
This includes artificial insemination, embryo transfer, *in vitro* fertilization, manipulation, and development of oocytes and cryopreservation.

The individual protocols used for human infertility treatment and domestic and wild mammal manipulation may vary in their objectives and in their detail but they are very similar in the mechanisms employed. Various general procedures will therefore be considered in turn.

6.1 INDUCTION OF FOLLICULOGENESIS, OESTRUS AND OVULATION

Depending on the species and objectives, these procedures may involve:

(a) The stimulation of normal preovulatory follicular development and ovulation rate (and accompanying oestrus if appropriate) in conditions in which such functions are otherwise impaired (during prepuberty, seasonal and lactational anovulation and in diverse forms of infertility).
(b) The stimulation of increased preovulatory follicular development and ovulation rate to increase litter size or to provide embryos for embryo transfer and twinning.
(c) The stimulation of increased preovulatory follicular development (without ovulation) to provide oocytes which can be collected for *in vitro* fertilization and early development.

The basic procedures (Figure 6.1), therefore, involve the manipulation of ovarian function in a controlled fashion to induce preovulatory follicular development and ovulation of one or more oocytes either to be fertilized *in vivo* and left *in situ* or flushed from the tract for freezing or transfer, *or* for removal for *in vitro* fertilization and development.

In all cases the method employed requires the administration of an appropriate dose of gonadotrophin or stimulation at some point in the pathway controlling gonadotrophin release.

6.1.1 The use of exogenous gonadotrophins

Such gonadotrophins may be derived from pituitary glands by extraction and purification, made by recombinant DNA technology, or pre-

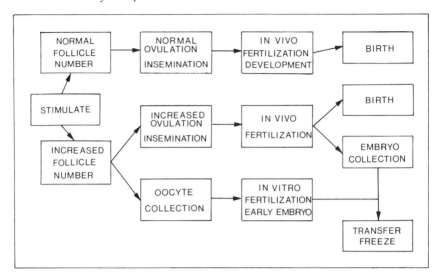

Fig. 6.1 General protocols for stimulating normal follicular development, and multiple follicular development with and without *in vitro* fertilization and embryo transfer.

pared from human menopausal urine or from the blood or urine of the pregnant mare or human, respectively (see Chapter 4, page 62). Administration of exogenous gonadotrophin results in recruitment and ovulation of more follicles than normal. There are at least three possible mechanisms of action. The increased concentration of gonadotrophins may either exceed the threshold for recruitment of the normal number of follicles, or it may lengthen the duration of gonadotrophin elevation, leading to more follicles reaching a recruitable stage during stimulation. The rescuing of follicles that have already begun atresia seems less likely. Stimulation with FSH is sustained to overcome 'dominant follicle' effects in those species in which this normally occurs. If the aim is to induce a single ovulation it may be possible to stimulate a cohort of follicles including a 'dominant' follicle and then to reduce FSH levels to allow the others to regress.

After the stimulation of preovulatory follicular development additional gonadotrophin stimulation is not usually required to induce ovulation, as oestrogen generated by follicle stimulation will induce an endogenous LH surge. In some circumstances, however, the timing of ovulation may be improved by exogenous LH administration, and patients with hypothalamo–pituitary failure will clearly require an ovulatory dose of LH probably in the form of hCG following follicular recruitment with FSH.

Examples of procedures used to induce superovulation in cattle with PMSG and ovulation in a woman with hMG are shown in Figures 6.2 and 6.3, respectively.

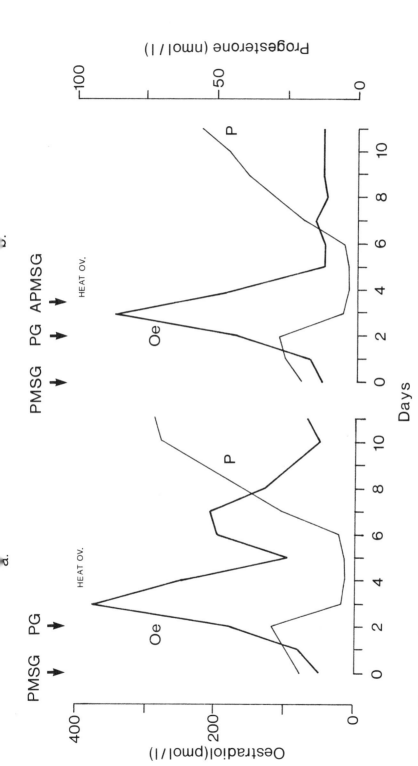

Fig. 6.2 Changes in plasma oestradiol and progesterone concentrations in cows injected with exogenous gonadotrophin (pregnant mares' serum gonadotrophin, PMSG) with and without the addition of an injection of an anti-PMSG antibody. (a) Treatment with PMSG only. (b) Treatment with PMSG followed by anti-PMSG (APMSG). Arrows on graph indicate times of administration of PMSG, PG (injection of prostaglandin $F_{2\alpha}$ analogue to induce luteolysis), and APMSG. Oe, oestradiol; P, progesterone; HEAT, time of oestrus; OV, time of ovulation. (*Data from:* Alfuraiji, M., Broadbent, P.J., Hutchinson, J.S.M., Atkinson, T., personal communication).

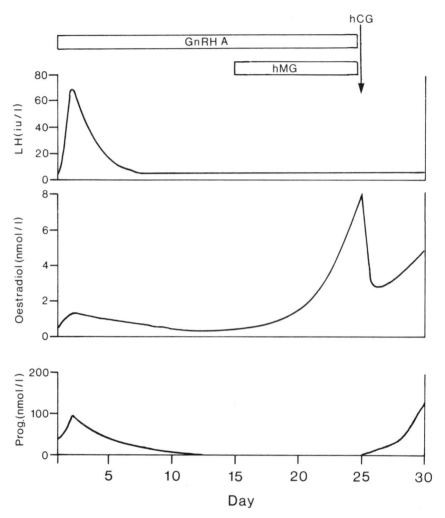

Fig. 6.3 Changes in plasma concentration of LH, oestradiol and progesterone in a woman treated with exogenous gonadotrophin to induce follicular development and ovulation, in conjunction with the use of gonadotrophin block with GnRH agonist. GnRH A, duration of GnRH agonist treatment; hMG, duration of human menopausal gonadotrophin treatment to induce follicular development; hCG, time of treatment with human chorionic gonadotrophin to induce ovulation. (*Data adapted from*: Fleming, R., Jamieson, M.E., Coutts, J.R.T. (1990) in *Clinical IVF Forum* (eds P.L. Matson, B.A. Lieberman) Manchester, University Press, p. 7).

Several major problems have arisen with the use of exogenous gonadotrophin to stimulate follicular development and ovulation in various species. First, there is in general a great variability of response between individuals and from one treatment to another within individuals. Differences in response have also been demonstrated between breeds and in different physiological states. For example, non-obese women respond to lower doses of gonadotrophin than do hypogonadotrophic individuals. Exogenous gonadotrophin stimulation may also result in an asynchrony in maturation rate among oocytes or among follicles and between oocytes and their follicles. Second, the ratio between FSH and LH in the preparation chosen for follicular stimulation is important. Preovulatory follicular stimulation requires both FSH and LH; hence preparations containing an optimal FSH:LH ratio are required for suprafollicular recruitment. Problems have arisen with preparations containing too much LH, which may affect thecal androgen production. This may impair oocyte development, producing lower ovulation rates and fewer good quality embryos. The endogenous production of LH may also be important: individuals with gonadotrophin failure may require more LH than those with normal pituitary gonadotrophic function. Preparations which contain a fixed ratio of FSH:LH are usually used in both human and animal studies. No account can be taken, therefore, of possible variations needed for the recruitment of ovarian follicles during their development. Different molecular forms of FSH and LH may be needed at different times of follicular development. The advent of pure recombinant FSH and LH with specific half-lives may allow for the use of more tailor-made treatments. Interference by endogenous gonadotrophins, during treatment with exogenous gonadotrophins, may also be important. Improved ovulation responses have been obtained by suppressing endogenous gonadotrophin secretion, particularly inappropriate LH surges, with GnRH agonists during gonadotrophin (and clomiphene) therapy (Figure 6.3).

Overstimulation can also be a major problem in gonadotrophin therapy, leading to multiple births or the more serious hyperstimulation syndrome in women, and to the development of cystic and secondary follicular development in cattle. The latter effect is a particular problem when long-acting preparations such as PMSG are used for follicular stimulation. The long-acting nature of PMSG is an asset in that only a single injection is needed (Figure 6.2), but it can lead to secondary follicular stimulation and an additional oestrogen rise after ovulation, with possible impaired early embryo development. Such secondary follicular stimulation and oestrogen rises can be prevented by the administration of monoclonal antibodies against PMSG, which 'mop up' the excess PMSG in the blood (Figure 6.2), but subsequent improvements in embryo quality have often been disappointing.

The ovarian response to exogenous gonadotrophins is related to the state of follicular development at the time of FSH treatment. Optimal responses are obtained when gonadotrophins are started around days 5–7 of the follicular phase in primates and late in the luteal phase of natural or artificially synchronized cycles in ruminants. In species which produce a dominant follicle (e.g. cow, monkey), gonadotrophins are given before final selection of the dominant follicle, after which other follicles may be undergoing atresia and not respond. Some studies in sheep and cattle have demonstrated an improved response to super-ovulation treatments when a priming dose of FSH is given early in the cycle prior to the FSH treatment. Such an early cycle peak of FSH has been demonstrated in untreated cycles.

Current ovulation induction regimens used in man or animals take little account of the importance of many of the endocrine, autocrine, and paracrine mechanisms other than gonadotrophins now known to be important in follicular development (see Chapter 2, page 33). If treatments are to be refined the roles of oestrogens and growth factors (IGF-I, TGFs, inhibin, activin) need to be defined. Some women who respond poorly to exogenous gonadotrophins show improved follicular responses after administration of growth hormone, suggesting perhaps an important role for IGF-I in some cases.

6.1.2   The stimulation of gonadotrophin secretion by GnRH or GnRH analogues

In order to achieve appropriate gonadotrophin secretion to stimulate follicular maturation, GnRH can be administered in either an episodic manner or by continuous infusion at low dosage. This can now be achieved by the use of continuous injection 'minipumps' that can be either implanted under the skin or taped on externally. Continuous stimulation with high levels of GnRH, however, causes inhibition rather than further stimulation. Such an action is currently being investigated for improving results of gonadotrophin therapy (see page 106). GnRH has also been used to alleviate specific infertility problems. Follicular cyst formation leading to nymphomania is a particular problem in dairy herds. Follicular ovulation induced by GnRH treatment can alleviate the problem.

6.1.3   Indirect stimulation of GnRH release

Other methods of inducing folliculogenesis involve a variety of techniques for stimulating the episodic release of GnRH.

Pheromones

Teaser rams have been used to stimulate oestrus and ovulation in ewes prior to the normal breeding season. This 'ram effect' operates *via* a pheromonal mechanism which involves a rapid release of LH in ewes exposed to rams (Figure 6.4). The ovulation induced, which is normally

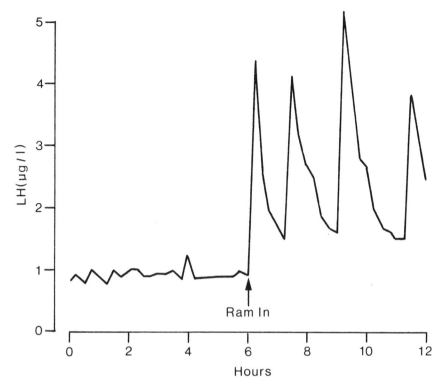

Fig. 6.4 The effect of exposure to a ram on the episodic output of LH in an ovariectomized/oestrogen-treated ewe. (*After*: Martin, G.B. (1984) *Biol Rev*, **59**, 53).

not accompanied by a heat period, is followed by either a short or normal length luteal phase, followed in turn by a normal ovulation and oestrus. A similar phenomenon in goats is caused by a related pheromone: ewes can be stimulated by teaser billy goats. An analogous situation occurs in young sows exposed to boars just prior to puberty. Early puberty is induced by a mechanism involving pheromonal transfer from the submaxillary gland of the boar to the vomeronasal organ of the sow.

Photoperiodic stimulation; melatonin

Photoperiod is the trigger of reproductive function in many seasonal species. Changes in photoperiod have been used to manipulate reproduction, but the need to provide light-controlled housing is a severe barrier to its practical application. The finding that melatonin secretion mediates the photoperiodic information (see Chapter 3, page 50) has led to the use of exogenous melatonin not only to advance but also to reverse the breeding season in highly seasonal breeds of sheep and goats, to increase ovulation rate at first oestrus, and to induce year-round ovarian cyclicity in red deer. Practical methods of administration, which include intra-rumen slow-release devices, ear implants, and vaginal sponges, have been developed.

Modification of feedback responses and other integrative mechanisms

Oestradiol administration can induce ovulation in some circumstances by the stimulation of an LH surge. Of more potential application has been the finding that active or passive immunization against both negative feedback steroids and inhibin (see Chapter 4, page 65), in some species will remove negative feedback influences present during the preovulatory period, and hence cause an increase in ovulation rate (Table 6.1). The administration of the steroid enzyme inhibitor, epostane (see also Chapter 4, page 64) during the luteal phase, prior to mating, also increases the ovulation rate in ewes. Such an affect may also act by changing gonadotrophin secretion by removing steroid negative feedback. In certain anovulatory women, particularly those with the Stein–Leventhal syndrome (see Chapter 3, page 54), in whom altered feedback mechanisms result in abnormal gonadotrophin release and polycystic ovaries, ovulation can be induced by an antioestrogen, clomiphene. Clomiphene appears to act by blocking oestrogen receptors in the hypothalamo–pituitary axis, allowing a normal sequence of FSH and LH release (Figure 6.5), and leading to a normal ovulation.

Lactation and suckling after parturition cause anovulation by mechanisms involving increased prolactin secretion or, in some cases at least, oxytocin release. In cattle, sheep, and pigs an opiate peptide mechanism is involved in the inhibition of GnRH release that occurs during the suckling period. Under such circumstances LH secretion can be induced by the administration of an opiate peptide receptor antagonist such as naloxone (Figure 6.6). In some women, infertility and anovulation are associated with increased plasma prolactin concentrations, often due to a microadenoma of the anterior pituitary gland. Such infertility can frequently be alleviated by the use of a dopamine agonist, bromocriptine, which probably acts by mimicking

Table 6.1 Ovulation rates in control and andro-stenedione immunized ewes

Control	Immunized
1.1 ± 0.1	1.6 ± 0.2

(*Data from*: Campbell, B.K., Scaramuzzi, R.J., Evans, G., Downing, J.A. (1991) *J. Reprod. Fertil.*, **91**, 655–66)

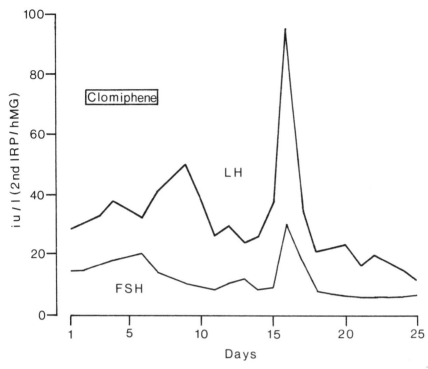

Fig. 6.5 The effect of clomiphene on gonadotrophin secretion in a woman with polycystic ovarian disease. The initiation of a 'follicular phase' rise in FSH is followed by a preovulatory gonadotrophin surge. (*Adapted from*: Ross, G.T. (1985) in *Williams Textbook of Endocrinology*, 7th Edn. (eds J.D. Wilson, D.W. Foster) Philadelphia, Saunders, p. 246).

dopamine, an endogenous inhibitor of prolactin. In addition to the manipulation of such hypothalamo–pituitary–ovarian mechanisms a variety of integrated intraovarian mechanisms (see Chapter 2, page 33) will no doubt be manipulatable. Such methods await development.

6.1.4 'Genetic' manipulation of ovulation

Some breeds of sheep have naturally high ovulation rates. Two mutations have been shown to affect ovulation rate in sheep, an autosomal mutation (Fec^B) in the Booroola and a mutation located on the X chromosome ($FecX^I$) in the Inverdale. Thus, in the Booroola, homozygotes (BB), heterozygotes (B+) and non carriers (++) have 5, 3–4, and 1–2 ovulations, respectively. The Fec^B gene appears to influence follicular growth before antrum formation, resulting in fewer granulosa cells at all phases of antral growth and more, smaller, follicles at ovulation. The mechanism involves either a modest increase in FSH secretion or

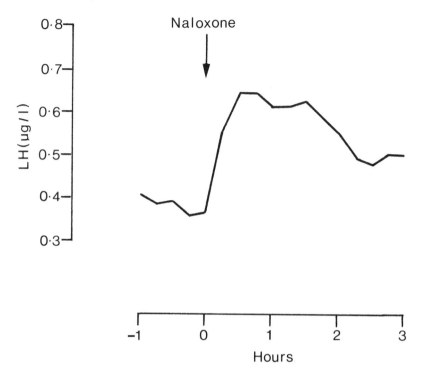

Fig. 6.6   The effect of naloxone in suckling sows. The LH response is measured after naloxone (1–4 mg/kg body weight) is given intravenously. (*Data from*: Barb, C.R., Kraeling, R.R., Rampacek, G.B., Whishart, C.S. (1986) *Biol Reprod.*, **35**, 370).

some yet unresolved mechanism. Ovulation rates in heterozygotes (I+) in the Inverdale are increased by about one egg per ewe. Homozygotes (II) are infertile, having streak ovaries containing primordial follicles but no growing follicles.

6.1.5   The manipulation of spermatogenesis and androgen secretion

There has been much less emphasis on manipulation of testicular function in the male than on follicular development and ovulation in the female. There also seems to have been a shift of focus from the potential treatment of male infertility by hormonal means to direct sperm manipulation (see page 127). Leydig cell function can be stimulated by exogenous LH-like gonadotrophins in all species studied: hCG has been frequently used because of its long half-life. The initiation and maintenance of spermatogenesis can be achieved with FSH and testosterone, respectively, but such treatments have been little used other than under experimental conditions. Hypogonadotrophic hypogonadism represents only a small percentage of human male infertility, but such individuals respond to both gonadotrophin treatment and pulsatile GnRH therapy. A very limited number of specific selected cases respond to naloxone and clomiphene. In men undergoing pituitary removal for treatment of, for example, a growth hormone-producing tumour causing acromegaly, spermatogenesis may be maintained by the injection of testosterone esters, if treatment is begun immediately. If there is a delay spermatogenesis must be reinitiated by administration of FSH.

6.1.6 Acupuncture

Acupuncture has been suggested for the control of ovarian steroidogenesis, post-partum anoestrus and cystic ovaries. Its effects may be mediated by paracrine or autocrine mechanisms within the glands, or perhaps *via* opiate-mediated mechanisms on the hypothalamo–pituitary axis.

6.2 PROGRAMMING AND SYNCHRONIZATION OF CYCLES

It is useful to be able to realign the events of the cycle in both man and animals. It may be an advantage for individual women or animals or groups of animals to undergo periovulatory events at a known or synchronized time, enabling timed artificial insemination, manipulations of the ovary for oocyte collection and superovulation, and timed or grouped births to be achieved. Progestogens have been used in a wide range of species and prostaglandin $F_{2\alpha}$ analogues have been used particularly in ruminants. GnRH analogues are being increasingly used and investigated.

6.2.1 Progestogens

Progesterone or a synthetic analogue is given either by mouth, by injection, by an impregnated vaginal device (Figure 6.7), or by an ear implant. An artificial luteal phase is induced and preovulatory follicular recruitment and ovulation are blocked by negative feedback. Certain commercial versions of vaginal devices are called PRIDs (progesterone releasing intravaginal devices) and CIDRs (controlled internal drug release dispensers). The use of such a device in animals is shown in Figure 6.8b: animals at any stage of the cycle when the device is inserted will have a prolonged luteal phase. Oestrus and ovulation will occur in a synchronized manner on removal of the device from all

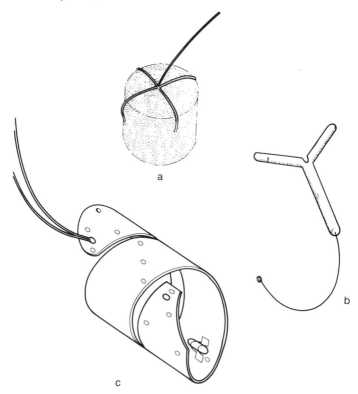

Fig. 6.7   Some vaginal progestogen devices. (a) Sponge, (b) CIDR, (c) PRID.

Fig. 6.8   The principles of cycle synchronization using either exogenous progestogen or a prostaglandin analogue. (a) Schematic representation of a ruminant oestrous cycle showing ovulation, corpus luteal activity and preovulatory follicular activity. (b) The use of progestogen to extend the luteal phase in four ewes which have ovulated a day apart. On withdrawal of the progestogen follicular development, oestrus, and ovulation will be synchronized in all four animals. This principle is applied to ewes which, in practice, ovulate over an approximately 16-day period. (c) Animals ovulating sequentially can be synchronized by injecting a prostaglandin analogue during the luteal phase when luteolysis can be induced. In practice to synchronize oestrus and ovulation in all animals during, for example a 21-day cycle in the cow, two injections are given 11 days apart (see text for details). Ov, ovulation; CL, corpus luteum; Foll, developing preovulatory follicle; PG, time of injection of prostaglandin analogue.

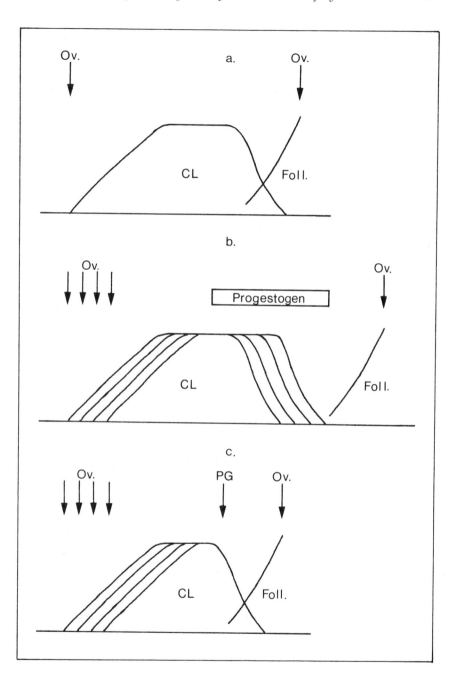

animals, as long as the device is left in place long enough for existing corpora lutea to regress or a luteolytic agent (such as oestrogen in cattle PRIDs) is added to the device to cause luteal regression. For use with artificial gonadotrophin stimulation, apart from convenience, such treatment may create a more homogeneous cohort of follicles and a lower incidence of spontaneous LH surges. A disadvantage, at least in women, is some evidence of ovarian insensitivity to gonadotrophins with the development of fewer preovulatory follicles.

6.2.2   PGF analogues

In species in which the corpus luteum of the non-pregnant oestrous cycle is regressed by $PGF_{2\alpha}$ and in which the corpus luteum is sensitive to such luteolytic action for half the length of the cycle (e.g. cattle, sheep, goats, but not pigs), several slowly metabolized analogues have been used to induce corpus luteal regression, allowing oestrus and ovulation synchronization within groups of individuals (Figure 6.8c). A standard procedure is to give two injections of the $PGF_{2\alpha}$ analogue half a cycle length apart to synchronize all animals in a group. In cattle, for example (with an approximately 21-day oestrous cycle) animals on days 5–17 of the cycle at the time of the first injection (bearing a corpus luteum sensitive to $PGF_{2\alpha}$) will undergo luteolysis, display oestrus about 3 days later and ovulate approximately 4 days later. At the time of the second injection, 11 days after the first, all these cows will be at approximately day 8 of the oestrous cycle and will respond again to the $PGF_{2\alpha}$. Cows between days 18 and 4 of the cycle would not have responded to the first injection, but will be between days 8 and 15 at the time of the second injection. All cows in a herd may, therefore, be expected to have a synchronized oestrus about 3 days after the second injection of the $PGF_{2\alpha}$ analogue, and ovulate approximately 4 days after.

6.2.3   GnRH analogues

GnRH agonists are being increasingly used for cycle programming. Upon administration there is a brief period of gonadotrophin stimulation. The pituitary receptors then become down-regulated (see Chapter 2, page 18) and there is a prolonged period of suppression while the treatment lasts, followed by recovery upon withdrawal. GnRH agonists may be particulary valuable for suppressing endogenous gonadotrophin production in cycling women or in animals undergoing exogenous gonadotrophin stimulation (see Figure 6.3, page 106). Patients with hypogonadotrophic anovulation may respond more uniformly to exogenous gonadotrophins than those with abnormal but active pituitary function, such as polycystic ovarian disease. The induction of a reversible hypophysectomy state in cycling women and animals undergoing gonadotrophin stimulation may prove advantageous by

removing inappropriate LH surges (causing possible premature luteinization and early ovulation) and by diminishing the variability of
response, by yielding a more synchronous follicular cohort for stimulation. Such treatments have already proved very valuable in women
(Figure 6.3, page 106).

6.3   MANIPULATION OF GAMETES AND THEIR TRANSFER BETWEEN INDIVIDUALS

A normal or an increased number of embryos may be obtained by
increasing the ovulation rate *in vivo* followed by natural mating or
artificial insemination. Reproductive events can also be increased and
timed by the use of various procedures for manipulations of the gametes
and their artificial transfer into individuals (e.g. *in vitro* fertilization
(IVF) and early embryo development, artificial insemination and embryo transfer and their variants, embryo manipulations, and preservation of gametes and embryos).

6.3.1   *In vitro* fertilization and early embryo development

Oocyte recovery

In vitro fertilization requires the collection of oocytes from ovarian
follicles, either at laparoscopy or after slaughter. In the latter case
oocytes may be obtained from unstimulated ovaries. In women, and
perhaps as procedures develop in other species also, oocytes are usually
obtained after stimulation with exogenous hormones (see page 103) to
ensure that several mature oocytes are available. Patients are monitored by ultrasound, and determination of oestradiol and/or LH levels
to enable collections to be made a few hours before ovulation. Oocytes
collected from antral follicles from animals at slaughter or *in vivo* from
unstimulated animals will normally be obtained from immature follicles
which may be either developing or undergoing atresia. Best *in vivo*
results are obtained with oocytes from non-atretic clear transluscent
follicles with an intact compact cumulus layer enclosing the oocyte.

Oocyte maturation

Oocytes collected from suitable immature follicles (see above) may be
matured *in vitro* in culture medium. A temperature above 37°C (39°C) is
optimal for several species (e.g. cow, sheep, goat). To induce full
maturation of the nucleus and cytoplasm, several factors, which may
be important during the latter phases of oocyte maturation within the
follicle, have been added to the incubation medium. These include

gonadotrophins, oestrogen, serum, and granulosa cells. An adequate and intact granulosa covering and co-culture with granulosa cells have been shown to be important for maturation, *in vitro* fertilization, and development to the blastocyst stage.

In vitro *fertilization of oocytes*

After evaluation (see Chapter 5, page 73) fresh or frozen sperm are washed in a suitable medium (which varies from species to species) to induce capacitation. Sperm are then placed in culture with oocytes. Temperatures a little above 37°C may again be critical for success. The next day the oocytes are examined for evidence of fertilization.

Development of the embryo

Continued incubation under controlled conditions leads to the development of preimplantation embryos, up to the blastocyst stage. Great difficulties have been encountered with this process. Specific modifications to the culture media are required at specific stages of development and for different species. Co-cultures with, for example, oviductal cells to mimic the early environment of the cleaving embryo, have also been developed. Such co-cultures may be either adding stimulatory factors to the medium (positive conditioning) or removing inhibiting factors (negative conditioning). At the end of the developmental, culture period preimplantation embryos may be either frozen for future use (perhaps for transportation to another venue) or transferred back to the supplier of the oocytes or to another individual (see page 119).

Efficiency of IVF and early embryo development techniques

After overcoming initial difficulties blastocysts have been produced. There is, however, generally a low efficiency of development of oocytes through to the blastocyst stage. This may be because very small 'errors' of technique have critical effects on early embryo development. Small differences in the temperature during fertilization, for example, may be crucial for subsequent success. There may also be a high incidence of genetic abnormality in the embryos. The rate of early embryo loss *in vivo* is high (see Chapter 3, page 55), much of it due to genetic abnormalities. An equal or even greater incidence of such abnormality may be expected in artificially derived oocytes.

6.3.2   The manipulation of oocytes and embryos

The increased availability of oocytes and embryos resulting from the application of superovulation and *in vitro* fertilization and development, and advances in biotechnology, has led to the development of a number of new techniques for increasing reproduction and genetic potential, including the production of transgenic animals and cloning.

Gene transfer

It is possible to transfer genes into a fertilized egg, allowing the replacement of a missing gene or the introduction of an exotic gene. Such methods offer not only possibilities for the introduction of valuable productivity traits into domestic animals, but the alleviation of genetic abnormalities, and the addition of genes involved in increased prolificacy or reproductive efficiency.

Cloning

The production of groups of large numbers of identical individuals is of great potential in domestic animal improvement and in the conservation of endangered species. Two different methods have been developed: embryo splitting and nuclear transfer. The former involves the bisection of embryos between the two-cell stage and the early blastocyst, to yield two embryos. Splitting into four and beyond becomes a limitation as the numbers of cells decrease. Split embryos survive freezing and thawing poorly. Nuclear transfer involves the transfer of nuclei from blastomeres or karyoplasts to enucleated oocytes. This overcomes the limitations of embryo splitting although the efficiency of the technique is still quite low. Embryo stem cells from the inner cell mass of the embryo retain their ability to divide in culture. Such cultured stem cells could be used as donors of nuclei for transfer to produce potentially very large clones. Such stem cells could also be genetically modified.

Other manipulations of embryos include sexing (see Chapter 5, page 98), freezing (see page 125) and artificial fertilization, in which a sperm is introduced through the wall of the zona pellucida (intrazonal fertilization) to overcome problems at IVF associated with either low sperm numbers or ineffective sperm–oocyte recognition.

6.3.3 Embryo transfer

This is a procedure in which preimplantation embryos, collected from one female (donor) or developed *in vitro*, are transferred to the same or another female (recipient). The first embryo was transferred between rabbits by Walter Heape a century ago. Transfers in small mammals were developed in the 1930s but it was not until the 1950s that major advances took place, with the transfer of pig and cattle embryos. The development of non-surgical transfer in the 1960s increased interest in

nd the first transfer of an IVF human embryo leading
rth took place in 1978. Irrespective of species or the
dy the methods employed have been very similar.
oviductal stage embryos is required surgical tech-
usually used; for the transfer of uterine stage embryos
..e has been a move towards procedures requiring progressively less
surgery and, if possible, transfer is performed *via* the cervix. Although
surgical transfer may be slightly more effective, non-surgical transfers
are clearly more practical. In all cases it is important that the donor
reproductive tract should be as nearly as possible at a synchronous
stage of development to that of the transferred embryo. The method
has allowed the exploitation of a wide range of techniques (IVF, frozen
embryos, twinning, multiple ovulation), the application varying with
the objectives and the species involved.

Transfers in women

Embryo transfers in women are usually derived from oocytes fertilized
and developed *in vitro* in an attempt to overcome infertility due to
blocked tubes or other conditions (Figure 6.9). The average success rate
varies between centres (12% to 25%), and the average 'take-home
baby' rate per treatment cycle is dramatically reduced with age, from
about 20% in the 25–34 age group to 7% in the 40–44 age group. The
procedure may involve oocyte donation or surrogacy. Oocyte donation
is now potentially possible for women with ovarian failure, due to
either a sex chromosome abnormality such as Turner's syndrome, or
premature ovarian failure which may either involve an early menopause
or oocyte loss due to ovarian ablation, chemotherapy or radiotherapy;
or those with normal ovarian cyclicity who either fail to respond to
other forms of infertility treatment or, being over 40, have a high risk
of conceiving a baby with Down's syndrome or some other genetic
abnormality. Patients with ovarian failure may require sequential
oestrogen and progesterone treatment to induce development of an
endometrium capable of allowing blastocyst implantation and maintaining early pregnancy. Surrogacy may, according to circumstances,
involve the transfer to a woman of an embryo derived from either her
own oocyte and a sperm from the potential father, or an oocyte and a
sperm from the potential parents, or perhaps totally unrelated gametes.

Transfer in other species

Embryo transfer has a major role in both domestic mammals and
endangered species, with the transfer of preimplantation embryos
developed either *in vivo* or *in vitro*, and which may be either fresh or
previously frozen. The technique has a particular role in obtaining

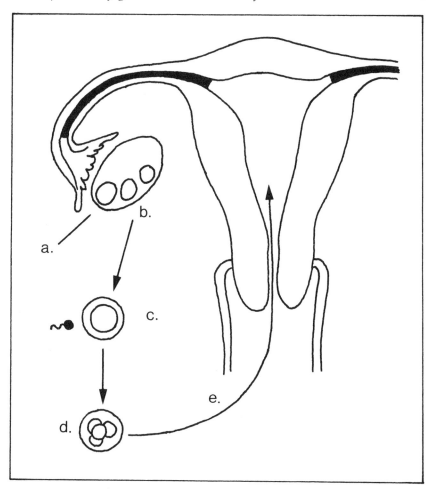

Fig. 6.9 *In vitro* fertilization and embryo transfer in women. After development of multiple follicles by gonadotrophin or other treatment (a), oocytes are removed (b) and fertilized *in vitro* (c). Early embryos are then developed *in vitro* (d) and transferred to the uterus (e).

increased numbers of offspring from either infertile, genetically high quality, rare, or endangered species. In the case of infertility it is important that this should be due to disease, injury or age and should not be genetic in origin. Treatment is likely to involve either superovulation or *in vitro* fertilization and development.

Multiple ovulation and embryo transfer (MOET) is a composite technique, usually including cycle synchronization, superovulation, artificial insemination, embryo recovery, and embryo transfer (Figure 6.10). Theoretical predictions have, however, exaggerated the potential increase in rates of genetic change available from the use of MOET,

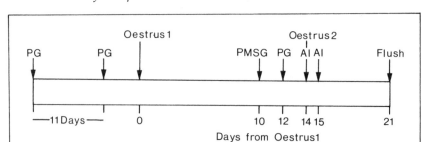

Fig. 6.10 An example of multiple ovulation and embryo transfer (MOET) in cattle. Two injections of prostaglandin analogue (PG) are given 11 days apart to synchronize cycles. Gonadotrophin (PMSG) is given on day 10 of the induced cycles followed by prostaglandin analogue to regress existing corpora lutea. Animals are artificially inseminated (AI) and embryos collected (Flush).

particularly in a small closed MOET nucleus scheme (i.e. involving a nucleus herd of elite males and females).

In endangered species it may be possible to use a more readily available related species as a surrogate mother. Transfers have been made for example between the Gaur cow and the domestic cow; from the Bongo antelope to the eland; and from Grant's zebra and the Przewalskii's horse to the domestic horse (Table 6.2).

Twinning of species which usually only have single offspring but which are capable of carrying twins can be achieved by embryo transfer. Transfer may be either of two embryos to a synchronized recipient or of one embryo to a mated synchronized recipient. The use of such techniques in cattle have problems due to the potential development of freemartins. Such problems can be overcome by the use of sexed embryos or sperm to ensure the transfer of two embryos of the same sex.

Table 6.2 Surgical transfer of embryos of Przewalskii's horse and Grant's zebra to domestic mares

	No. of embryos transferred	No. of pregnancies established	No. of pregnancies to term	No. of live foals
Przewalskii's horse to domestic horse	9	7	4	3
Grant's zebra[a] to domestic horse	5	3	2	1

[a] Grant's zebra is not endangered but may be regarded as a model for an endangered species transfer.
(*From*: Summers, P.M., Shepherd, A.M., Hodges, J.K., *et al.* (1987) *J. Reprod. Fertil.*, **80**, 13–20)

6.3.4 The value of methods involving multiple follicular development/ovulation, *in vitro* fertilization and embryo transfer

These techniques enable a number of infertility problems to be alleviated, increased reproductive efficiency and genetic improvement to be achieved, and endangered species to be conserved.

Infertility may be treated in women with blocked fallopian tubes, and in some with unexplained infertility. An example of the latter is a couple in which the wife has regular cycles and the husband has a satisfactory semen assessment, so that fertilization is likely to occur. The method also allows various forms of surrogate motherhood. Genetic problems may be checked and potentially alleviated, and embryos may be sexed.

Increased prolificacy of desirable genotypes of domestic and endangered or wild species may be achieved. Embryo banks can be set up and embryos may be transported conveniently and free of disease. Genetic improvement can be achieved, since by increasing the reproductive rate of the females the intensity of selection amongst them can be increased. In addition, the methods allow various manipulations of the oocytes and embryos for improved prolificacy and genetic potential.

6.3.5 Artificial insemination (AI)

This involves the artificial transfer of sperm, either fresh or after freezing, used either *in vivo* or in connection with IVF. This very old technique was probably first used in 1780 and became extensively used in horses and cattle from around 1900. The technique is now widely used throughout the world for a variety of domestic animals and endangered species, and in man, in connection with the use of high quality sperm for genetic upgrading, for conservation of endangered and other species, or for AI by husband or donor in man. Ejaculations may be collected by means of an artificial vagina (Figure 6.11), manual massage, electro-ejaculation, with the aid of teaser animals, or following sexual stimulation. After collection and evaluation the sample may be concentrated for motile sperm by a variety of techniques (see Chapter 5, page 77) and/or 'extended' with various isotonic buffer solutions containing an energy source, protective protein (skimmed milk, egg yolk), and antibiotics. Sperm are frequently stored by freezing (see page 125). Artificial insemination (perhaps by placement directly into the uterus) has proved to be a convenient and reliable procedure. Used in connection with freezing it has had a great impact on genetic improvement, disease control, elimination of dangerous animals from farms, reproductive management, wild species captive breeding programmes, and the setting up of sperm banks. It has been used in seasonal breeders, for the treatment of certain types of male

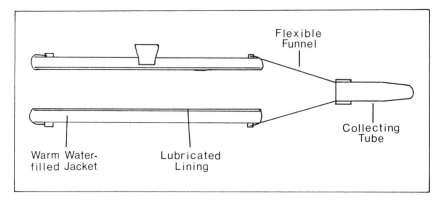

Fig. 6.11   Diagram of an artificial vagina. (*After*: Hammond, J., Mason, I.L., Robinson, T.J. (1971) *Hammond's Farm Animals*, 4th Edn. London, Arnold, p. 60).

infertility, and in men about to undergo surgical procedures that may adversely affect fertility (e.g. removal of the anterior pituitary gland).

6.3.6   Alphabetical techniques

In addition to IVF and AI, a large number of related techniques involving the transfer of gametes between individuals or in connection with *in vitro* techniques have been developed for the treatment of human infertility and for increasing reproductive efficiency in animals. A wide

Table 6.3   Some 'alphabetical' techniques used for gametes and embryos

Method	Acronym
Embryo transfer	ET
In vitro fertilization/embryo transfer	IVF/ET
Artificial insemination	AI
By donor	AID
By husband	AIH
Intrauterine	IUI
Intracervical	ICI
Intraperitoneal	IPI
Intrazonal fertilization	IZF
Gamete intrafallopian transfer	GIFT
Peritoneal oocyte and sperm transfer	POST
Transcervical intrafallopian transfer	TIFT
Transuterine fallopian transfer	TUFT
Zygote intrafallopian transfer	ZIFT

array of alphabetical acronyms have been coined to describe these processes (Table 6.3).

Gamete intrafallopian transfer (GIFT) is one of the techniques most used in man. This involves the production of follicles containing mature oocytes by the administration of gonadotrophins or other techniques (see pages 103–8). The oocytes are then collected, selected, and transferred to the ampullary region of the fallopian tube along with sperm. The method depends on the ability of the fallopian tube to act as the site of both sperm capacitation and fertilization. Unlike IVF, this latter process cannot be seen with the GIFT technique. Compared to IVF, however, GIFT has potential advantages: fertilization occurs at the natural site, gametes are exposed to a minimum of *in vitro* technique, and early embryos develop *in vivo*. The use of GIFT has been reported in the monkey and cattle but the site of deposition of the gametes, the fallopian tube, may limit the practical use of the technique for many other non-human species.

6.3.7   Preservation of gametes and embryos

Short-term procedures

Preimplantation embryos enclosed in the zona pellucida may be handled and evaluated fairly easily, and simple protein-containing buffered saline is an appropriate medium. Fresh and extended semen are used for artificial insemination and the latter may be transported, in some cases, in appropriate media. There is some evidence that *in vitro* life may be extended by certain additives which may both aid in the maintenance of motility of sperm and act to aid sperm transport within the female tract.

Long-term procedures – freezing

Methods for freezing semen were developed by Polge, Smith and Parkes in 1949. The freezing of sperm and embryos has greatly aided the development and use of manipulative techniques. Banks of sperm and embryos can be built up from high quality domestic animals and endangered species for a variety of genetic improvement, prolificacy increasing, and conservation programmes. Detailed progeny testing and widespread transportation are possible. In man, embryos for transfer to surrogates and spare embryos collected from infertile women may be preserved for another attempt at embryo transfer or for other procedures. Sperm may be preserved for use in artifical insemination by donor (AID) or artificial insemination by husband (AIH) when the husband has poor quality sperm. This may also be used in an IVF programme or prior to surgery that may leave the man infertile. Both sperm and embryo freezing are now established routines in various assisted conception programmes in both animals and man. Despite extensive investigation and use over 45 years, however, methods for sperm freezing vary, and the need for additives other than a cryopreservative is controversial. In general, fertilized oocytes, preimplantation stage embryos, and sperm have been preserved by initial slow cooling, in the presence of a cryoprotectant agent such as glycerol or dimethylsulphoxide to prevent ice crystals from destroying or damaging the cells. Rapid cooling and storage in liquid nitrogen is then undertaken. Success rates are high, though there is some loss of quality due to damage during the freezing and thawing processes. Cryopreservation of unfertilized oocytes has been less successful. Reduced fertilization rates and increased polyspermy have been reported, probably due to the deleterious effects of freezing upon the zona pellucida. An alternative approach used for embryo preservation is vitrification, in which the cells are osmotically dehydrated prior to cooling by controlled equilibration in a highly concentrated solution of cryoprotectant.

6.4 CONTROL OF EMBRYO LOSS

All species show a high incidence of embryo loss. Some embryos may be lost due to congenital abnormalities; other losses may involve inadequate maternal recognition or maintenance of pregnancy, and these are potentially preventable. The administration of additional progesterone may have a role by improving embryo development, the uterine environment and endometrium development, and/or reducing the normal factors inducing luteolysis. In species that undergo active luteolysis such mechanisms may include induction of uterine oxytocin receptors by follicular oestrogens. The luteinization of such follicles with GnRH analogues may prevent embryo loss. Administration of maternal recognition factors may also be effective in some cases. Recombinant interferon-α_1 1 mimics the antiluteolytic action of the trophoblast protein produced by the conceptus in ruminants. The use of recombinant interferons for the prevention of embryo loss may, however, be limited.

The dangers of administering any compound during pregnancy to prevent embryo loss is that it may prevent the loss of abnormal embryos or even induce congenital abnormalities. The latter possibility has been recognized for at least 30 years. At that time the synthetic oestrogen, diethylstilboestrol (see Figure 4.3, page 64) was administered to prevent miscarriage in some women. Sadly many girls born to mothers so treated developed vaginal adenosis and carcinoma at puberty, when the reproductive tract again came under the influence of oestrogens.

6.5 TREATMENT OF HUMAN INFERTILITY

In man potentially diagnosable and treatable causes of infertility include ovulatory disorders (including hyperprolactinaemia), tubal and peritoneal factors, endometriosis, implantation failure, embryo loss, abnormalities of sperm–mucus interaction or capacitation and fertilization problems, and inadequate sperm production for various reasons. The highest chance of conception probably occurs amongst couples in whom anovulation is the only abnormality and in women under 35. Ovulatory problems associated with a thyroid disorder or raised prolactin levels may be treated by correcting the thyroid disorder or lowering prolactin with bromocriptine (see page 110). Deficiency of gonadotrophin can be overcome by clomiphene, FSH and hMG, or pulsatile GnRH. Polycystic ovarian disease may respond poorly to clomiphene or hMG: pure FSH or a GnRH agonist with hMG may be required. The key to treatment of this disease may be the understanding of the heterogeneity of the syndrome and its careful diagnosis (see Chapter 3, page 54). Treatment with a specific opiate peptide receptor antagonist (naltrexone) can produce follicular maturation and ovulation in some women with hypothalamic amenorrhoea. Tubal blockage may be treated by tubal clearance or microsurgery and increasingly by IVF and GIFT. Clearly, at least one healthy fallopian tube is necessary for the latter. Failure of implantation and embryo development is difficult to treat, despite the fact that an inadequate luteal phase, resulting in an inadequate endometrium, may be involved. Potential embryo loss due to genetic abnormalities may be increasingly diagnosed in IVF embryos, and only potentially viable embryos may be 'selected' for transfer. No special therapy is effective for the treatment of sperm–cervical mucus abnormalities. By-passing the problem by using IUI, IVF or GIFT is possible. Defective capacitation and fertilization cause infertility in a number of cases where the cause is otherwise unknown.

In the male, as in the female, primary gonadal failure is irreversible. Hypogonadotrophic hypogonadism may respond to gonadotrophins or pulsatile GnRH. Suppression of sperm antibodies with adrenal steroids may be effective but there are risks of side-effects from such therapy. AIH can only be undertaken if adequate numbers of good quality sperm are available: sperm may be pooled before the procedure. Intrauterine insemination, GIFT, and IVF may be useful in some cases of male infertility. Some would regard GIFT as the treatment of choice for such unexplained infertility. *In vitro* fertilization may be linked with drilling or partial dissection of the zona pellucida and microinjection of sperm into oocytes. Epididymal sperm may be obtained for use in such techniques if the male ducts are blocked. Although pregnancy may result from IUI, GIFT or IVF, only with IVF can the fertilizing capacity of sperm be directly assessed.

6.6 THE CONTROL OF PARTURITION

In man, and in other species, problems which arise towards the end of pregnancy can be overcome if premature parturition can be induced in a controlled and non-traumatic way. In pigs and perhaps ewes it may also be convenient to achieve batch farrowing or lambing in animals which are predicted to give birth a few days apart. There is also the possible complication that a normal delivery of offspring may be followed by a retention of the placenta.

The hormonal cascade which causes parturition differs in different species (see Chapter 2, page 40); methods for the induction of labour at term also differ. In species such as cow, pig, and goat, in which the corpus luteum is the primary source of progesterone late in pregnancy, $PGF_{2\alpha}$ is effective, causing luteolysis; maternally administered glucocorticoids are less reliable. In cattle, dystocia is associated with the use of both $PGF_{2\alpha}$ and glucocorticoids, but the antiprogestogen RU486 (see Figure 4.4, page 65) has been used successfully to induce parturition. In sheep, which have a placental source of progesterone, $PGF_{2\alpha}$ is ineffective unless labour is imminent, but maternal glucocorticoids and a progesterone synthesis inhibitor (epostane, see Figure 4.4, page 65) are effective. Oxytocin is frequently the choice in man: prostaglandin is no more effective than oxytocin and has a narrow therapeutic range, with gastrointestinal side-effects.

6.7 SEX HORMONE THERAPY OTHER THAN TO ACHIEVE FERTILITY OR INFERTILITY

In addition to their use in increasing or assisting fertility or inhibiting fertility (see Chapter 7), hormones (particularly sex steroids) have been used to replace or mimic gonadal hormone production in a number of conditions.

6.7.1 Man

In patients with primary hypogonadism due to a genetic condition (e.g. Turner's syndrome) or gonadal removal, steroid replacement may produce an acceptable development of secondary sexual characteristics, and induce or maintain libido in men. Oestrogens relieve menopausal symptoms such as hot flushes and prevent the onset of osteoporosis in post-menopausal women. Initially, preparations containing conjugated equine oestrogens were used for such hormone replacement therapy (HRT). Oestrogens, which are most easily converted to oestrone, are now recommended. These are often administered as patches to avoid a first pass through the liver, thus reducing metabolic side-effects. A progestogen is given with the oestrogen to women with an intact uterus. Worries exist concerning a possible link between oestrogen therapy and cancer. The evidence is not conclusive. Some increased risk of cancer has been associated with 10 or more years of treatment with oestrogen alone, but combined oestrogen and progestogen treatment has been favoured recently. There is also unconfirmed evidence that oestrogen therapy may protect against breast cancer.

6.7.2 Animals

Appropriate steroid therapy may be given to male and female animals to induce or enhance male libido and induce oestrous behaviour, respectively. A variety of teaser animals have been used to check oestrous behaviour and the incidence of mating (see Chapter 5, page 70), including androgen-treated females, castrated males, and oestrogen-treated females.

6.8 POSSIBLE PROBLEMS ASSOCIATED WITH THE USE OF STIMULATING REPRODUCTIVE TECHNOLOGIES

Although AI and ET have great potential for increasing reproductive and genetic potential in animals, it seems unlikely that full potential can be achieved by these means alone. A combination of techniques with other biotechnological approaches (e.g. oocyte maturation, IVF, cloning, sexing, and gene transfer) may be required, each of which increases the potential complexity of the method.

It is also important to recognize that it is not possible to improve fertility or prolificacy of man or animals simply by applying a new technology. An inappropriate technology or treatment for infertility used without appropriate diagnosis is unlikely to succeed. Similarly transfer of embryos, for example into a herd with a high rate of embryo loss due to poor management, will not solve these problems.

There are worries that increased manipulations of the reproductive system and of gametes and embryos may increase the incidence of malformations, or may at least be associated with a constant rate of loss. Is there perhaps a limit that cannot be overcome? Do only a certain number of oocytes have the ability for normal development? There are also worries that the increased use of new technologies in animals will lead to increased inbreeding. A lack of genetic diversity may have important effects on all characteristics, but may also lead to reproductive abnormalities, including cryptorchidism, low libido, and follicular cysts. Domestic animals have been artificially selected for certain reproductive traits. Males, for example, have been selected for sperm production rather than for traits required in the wild, such as acceptability to females and fighting ability. The use of new technology (e.g. AI, IVF) is likely to increase such artificial selection. This is perhaps unimportant for domestic species, but may be important when considering methods to be used in conservation programmes for endangered species.

In man, a major problem associated with various methods of assisted conception, including induction of ovulation, IVF, and GIFT, is an increase in the incidence of multiple births. Apart from problems associated with feeding and upbringing, there may be problems associated with pregnancy and parturition. Perhaps half of all sets of quadruplets are born before 32 weeks gestation; two-thirds are born by caesarean section. Over one-half spend 1 month or more in intensive care. The conservative use of conception assistance procedures should not lead to such problems, although more than one embryo is usually replaced to increase the chances of success. Selective reduction in numbers may be achieved by selective feticide. Though many may support abortion of a fetus with Down's syndrome and other conditions, the use of abortion because of too many potential brothers and sisters will have less support. The use of ovulation inducing drugs and reconstructive tubal surgery also carries an increased risk of ectopic pregnancies.

Ovarian hyperstimulation syndrome is probably the most serious potential complication of gonadotrophin therapy in women. This arises about 1 week after the administration of the ovulatory dose of hCG and causes marked ovarian enlargement and increased vascularity.

CHAPTER 7

Control of fertility: contraception

There are potential needs to control fertility in human society, and in domestic and wild mammal populations. It has long been the practice for long distance camel drovers to insert an apricot stone into the vagina of their camels to prevent impregnation during very long journeys. The prevention of unwanted babies dates back over 4000 years with successive Egyptian records of an abortifacient, vaginal pastes, and medical tampons. Modern contraceptives probably date from the use of a linen sheath in 1564 to protect against syphilis. Widespread use dates from the vulcanization of rubber around 1840, from which condoms and cervical caps were made. Intrauterine devices date from the early 1900s and steroidal contraceptive tablets (often called pills) from the late 1950s.

In human society one contribution to the alleviating of problems associated with population expansion is to have readily available, safe, effective, reversible, and acceptable methods of contraception. These latter methods also serve the more personal needs for family planning, to enable delays in and the spacing of children. Probably half a million women die every year due to complications of pregnancy, the majority of whom are in the less developed world. Additional women suffer ill health as a result of pregnancy complications. Prevention of unwanted pregnancies and decreasing the incidence of unsafe abortions could significantly alleviate these problems.

There are many needs for methods to limit reproduction in animals. Such procedures greatly facilitate the husbandry of mixed sex groups of growing domestic livestock. Periodic or complete reproductive constraint is useful for urban pets, working dogs, racing greyhounds, horses, and camels. There is also a major need to control stray urban dogs, escaped feral species (e.g. mink, coypu) and introduced (e.g. rabbits in Australia) or wild pest (e.g. rats) species. Reproductive control can have a major contribution toward these problems. Reproductive constraints may also be important for the management of certain wild and endangered species, particularly in the increasingly common situations where the size and nature of the habitat has been restricted

Table 7.1 Established and promising methods for human contraception (including abortion and sterilization)

Natural methods; lactation, abstinence, rhythm methods
Condoms, female condoms, caps, diaphragms, sponges
Sterilization: male/female surgery, injections etc.
Intrauterine devices: copper devices, medicated devices
Steroidal methods; female oral contraceptives, implants, injectables, medicated vaginal rings and IUDs, male methods
Abortion: RU486, postovulatory methods
Vaccination: against pregnancy, other methods
GnRH analogues, gossypol, melatonin
Antibodies against components of spermatozoa/oocyte recognition

by man's incursions. Artifical control of the size of a herd may be undertaken far more selectively and less traumatically by contraceptive methods than by culling.

Much research effort has been directed at producing human contraceptives. The same methods in general apply for use in other species. The possible loci of action for potential contraceptive methods have been discussed in Chapter 4. The aim has been to achieve safe, effective, reversible and acceptable methods with no side-effects. Table 7.1 lists the more traditional methods, and some newer methods under development. The pattern of contraceptive use greatly changed with

Table 7.2 World established use of contraceptive methods (1985)

Methods	Users $(\times 10^6)$
Traditional	58
Rhythm	16
Withdrawal	21
Condoms	38
Vaginal	6
Sterilization	155
Female	108
Male	47
Intrauterine devices	80
Hormonal	61
Oral	55
Injectable	6
Total	398

(*Data from*: Mauldin, W.P., Ross, J.A. (1989) in *Demographic and Programmatic Consequences of Contraceptive Innovations* (eds S.J. Segal, A.O. Tsui, S.M. Rogers) NY, Plenum Press, p. 34)

the introduction of 'the pill' and modern IUDs. The numbers of users of established methods in 1985 are given in Table 7.2. The different methods will be discussed in turn.

7.1 PHYSICAL AND PSYCHOLOGICAL BARRIERS

7.1.1 Abstinence

This plays an important role in the control of fertility whether it is practiced on social or religious grounds, or because of a fear of pregnancy. 'No' is the most effective contraceptive method. Periodic abstinence has been widely practiced, particularly by those with a religious or other objection to physical or chemical contraceptive methods. During a menstrual cycle there is only a short period, around ovulation, when conception is possible. The so-called natural or symptothermal methods involve the woman being able to recognize this fertile time. Ovulation detection (using basal body temperature, cervical mucus examination or a combination of methods) is, however, difficult (see Chapter 5, page 93) and cycle lengths, particularly the follicular phase, are variable. Hence 'safe period' predictions are unreliable. It is also possible, though rare, for ovulation to be induced by intercourse.

7.1.2 Interrupted coitus

Such methods, which include coitus interruptus or withdrawal prior to ejaculation and coitus reservatus, in which the man avoids an orgasm, are effective. Pre-ejaculatory fluid is unlikely to contain spermatozoa, at least in numbers sufficient to effect fertilization. Another related method, coitus saxonicus, in which pressure on the perineum at the time of ejaculation causes ejaculation into the bladder, is also effective. The idea, however, that avoidance of an orgasm by the woman is effective for avoiding pregnancy is without substance.

7.1.3 Raising the scrotal temperature

Spermatogenesis only occurs normally at temperatures below the abdominal temperature (see Chapter 3, page 50). Scrotal warming or wearing of tight underpants, to continually hold the testis close to the abdomen, are, however, ineffective methods of contraception.

7.1.4 Vaginal douching and the use of vaginal spermicides

The douching of the vagina with a salt or other solution, is ineffective in preventing conception. The use of spermicides, although fairly ineffective when used alone, may be valuable for the supplementation of other vaginal methods employing caps, diaphragms, or sponges.

7.1.5   Condoms

A male-worn condom or sheath is a very widely used contraceptive method. They are clearly almost without adverse or side-effects, other than a reduced contact sensitivity for the user. Condom use also effectively reduces the possibility of venereal disease transmission and has been strongly advocated for the prevention of human immunodeficiency virus (HIV) infection causing acquired immune deficiency syndrome (AIDS), for which medicated antiviral condoms have been introduced (along with female condoms, see below). The effectiveness of condoms is difficult to judge as they are frequently not obtained from clinics and no follow-up is carried out. Condoms are, however, now produced with a very high degree of quality control; hence failure is probably mainly due to incorrect use.

7.1.6   Diaphragms, caps and sponges

A variety of caps, diaphragms and sponges (Figure 7.1) fitted in the vagina or over the cervix by the woman have been employed in combination with spermicides (and in future also with antiviral agents). More recently a female condom (vaginal sheath), which lines the vagina and covers some of the labia, has been devised.

With the introduction of oral contraceptives (see page 140) many women changed to these from using caps or diaphragms. The effectiveness of the latter depends on the motivation of the woman, and many perfect the method over time. Failure rates therefore may be high in some women but low in others after regular use (see Table 7.5, page 148).

7.1.7   Voluntary male and female sterilization

The development of techniques enabling safe and rapid operations to be performed (sometimes in outpatient clinics), the possible use of methods employing injections into the vas in men, and possible non-surgical approaches in women (such as using transcervical quinacrine pellets) have all led to increased interest in and use of sterilization. It may be the most common form of fertility control worldwide (Table 7.2), and its use is likely to increase. The attitude of potential users towards the operation is important. Sterilization was often equated with gonadectomy, particularly in men, and hence associated with a potential loss of libido, rather than with the occlusion or partial removal of the vas or fallopian tube to act as a barrier to gamete trans-

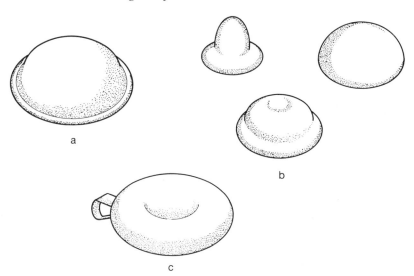

Fig. 7.1 (a) Vaginal diaphragm, (b) cervical caps, (c) sponge. (*From*: Fathalla, M.F., Rosenfield, A., Indriso, A. (1990) *FIGO Manual of Human Reproduction* Vol. 2, Carnforth, Parthenon, pp. 115, 119, 123).

port. Efforts to develop temporary or absolutely reversible methods have been unsuccessful. Though reversal may be possible in some cases, procedures should be entered into by the user as though they are irreversible. Fatalities and side-effects are rare. In men there may be some scrotal tenderness and migrating clips may be a complication, but suggestions that an increased leakage of sperm into the system could lead to the development of autoimmune disease seem unfounded. Although ectopic pregnancies occur in about 1 in 15,000 sterilized women, every sterilization preventing an unwanted pregnancy lowers the maternal death rate.

7.2 PREGNANCY TERMINATION – ABORTION

Legal and illegal abortion has been and remains an important method of 'family planning'. It is also probably the most controversial method for various ethical, religious, social, and biological reasons. Spontaneous abortion probably occurs in about half of all pregnancies, during which the majority of fetuses with gross congenital abnormalities (those not compatible with life) are lost. Some consider that the abortion of fetuses with other major abnormalities (e.g. Down's syndrome and spina bifida, diagnosed by amniocentesis or transcervical chorionic villus sampling), or termination of pregnancies with associated medical or psychological risks to the mother or following incest or rape is acceptable, if performed early. For others, religious and other reasons make any form of induced abortion unacceptable. Illegal abortion (often undertaken using non-professional or 'hopeful' mechanical, chemical, or 'knitting needle' surgical methods) always carries a high risk compared with medically approved methods. In general, abortion should be performed as early as possible: it becomes more dangerous as pregnancy proceeds.

7.2.1   Surgical methods

So-called menstrual extraction, in which the embryo and endometrium are sucked out, may be used within 2 weeks of a missed period. After 6–12 weeks a dilatation and curettage (D and C) is commonly employed and after 3–9 weeks a vacuum aspiration is often used. Later in pregnancy the use of either intra-amniotic hypertonic saline or fetal removal is necessary.

7.2.2   Hormonal methods

Prostaglandin analogues given by vaginal suppository or injection successfully terminate very early pregnancies, but the side-effects, especially nausea, may be severe. From 15–20 weeks prostaglandins are administered by injection into the amniotic fluid or between the fetal membranes and the uterine wall.

Steroids and steroid analogues have been used during the preimplantation stages to prevent pregnancy. These so-called postcoital steroid methods (see page 140) have been used particularly following rape. The side-effects may be severe.

7.2.3   Antiprogestational compounds

A short interruption of progesterone activity is sufficient to stop early pregnancy. Abortion can therefore be induced by substances that block progesterone synthesis, neutralize circulating progesterone, or block progesterone receptor binding (anti-progestogens).

Progesterone synthesis inhibitors

Several compounds have been synthesized which inhibit the activity of 3β-hydroxysteroid dehydrogenase, the enzyme involved in the conversion of pregnenolone to progesterone during progesterone synthesis (see Figure 2.6, page 16). Such compounds include trilostane, azastene, and epostane (see Figure 4.4, Chapter 4), the last being the most potent. Epostane induces premature labour in sheep and will terminate pregnancy in primates. In pregnant women epostane lowers blood progesterone concentrations, but pregnancy termination is usually only achieved after repeated administration over several days. Side-effects, especially nausea, are more severe than following administration of antiprogestogen compounds such as RU486 (see below). In addition, as blockers of progesterone synthesis, 3β-hydroxysteroid dehydrogenase inhibitors may block adrenal function (see Figure 2.6, page 16). This does not seem to occur, however, in primates at doses that are antiprogestational.

Antiprogestogen compounds

Compounds which exhibit high affinity binding to progesterone receptors in progesterone target cells, and thus prevent endogenous progesterone from binding to the receptor, but which have no progestational action, are termed antiprogestogens (see Chapter 2, page 18). A potentially useful group of compounds in this category comprises the 19-norsteroids (steroids lacking the methyl group at C10) with a *p*-(dimethylamino) phenyl group in the 11β position. The structure of one such compound (mifepristone) is shown in Figure 4.4 (page 65). Such compounds have been shown to be active at all stages of pregnancy, preventing pregnancy establishment, terminating both early and late pregnancy, and inducing parturition. Complete abortion follows a single dose of mifepristone (RU486) in about 80% of pregnancies of less than 42 days gestation. In conjunction with a small dose of prostaglandin given 36 h later, RU486 is over 95% successful up to 49 days and may be satisfactory for up to 67 days. In about 3% of patients abortion is incomplete and in about 1% there is no effect. The method seems, therefore, to be highly effective but medical supervision is mandatory. There are no side-effects directly attributable to the antiprogestogen. Blood loss is comparable to that following vacuum aspiration. The 1% of patients who are unaffected require surgical termination.

7.2.4 Anti-pregnancy vaccines

Several experimental programmes are working towards a vaccine to prevent pregnancy. Different potential target antigens have been suggested or identified, including sperm and zona pellucida (see page 148), cumulus oophorus, pregnancy hormones, and proteins of the female reproductive tract. A frequent approach has been to aim to interrupt the actions of hCG in the establishment and maintenance of early pregnancy. This can be achieved by either actively (direct immunization) or passively (using prepared antisera) immunizing women against hCG.

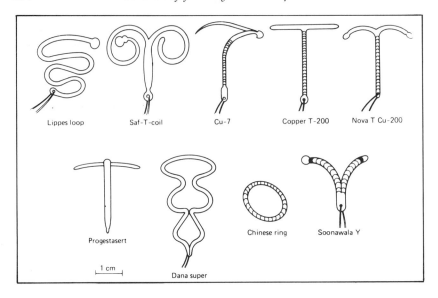

Lippes loop   Saf-T-coil   Cu-7   Copper T-200   Nova T Cu-200

Progestasert   Dana super   Chinese ring   Soonawala Y

1 cm

Fig. 7.2 Some IUDs in current use. (*From*: Jones, R.E. (1991) *Human Reproductive Biology*, NY, Academic Press, p. 314).

It has been possible to enhance the antigenicity of hCG by linking it to a carrier protein. One approach has been to use the β subunit of hCG with tetanus toxoid as the carrier. Non-uniformity of response and the possible cross-reactivity of the antisera generated with pituitary LH have been problems. Avoidance of cross-reactivity has been attempted by using only a specific portion of the hCG β subunit not present in human LH.

7.3   INTRAUTERINE DEVICES

The use of intrauterine devices for preventing pregnancy has a long history, as almost any foreign body placed in the uterus will prevent pregnancy. Such devices are useful for women who do not desire permanent sterilization or for whom steroidal contraceptives are contraindicated (e.g. smokers, see page 143). A wide range of devices has been tried in an effort to produce an effective product (Figure 7.2). A second generation of devices has been produced which incorporate copper (copper-coated), or slow release progestogens.

7.3.1   Mode of action and effectiveness

The precise mode of action of IUDs is not clear. They probably act by inducing mild chronic inflammatory responses and causing a migration of leucocytes into the uterine lumen; these phagocytose sperm and blastocysts. Larger devices are generally more effective. The copper ions in some devices may be both spermicidal and/or embryotoxic, and may increase the inflammatory reaction. Progestogen-containing devices may aid leucocyte invasion and induce the production of a cervical mucus hostile to sperm. IUDs have a 1–5% failure rate; the progestogen-containing devices may be even more effective.

7.3.2 Contraindications and complications

Intrauterine devices are contraindicated if pregnancy is suspected, and in patients with gynaecological malignancy or general infection. They should probably be avoided in those with a history of pelvic inflammatory disease (PID) or sexually transmitted disease. IUDs may cause bleeding and/or spotting due to compression of the endometrial lining (or due to the copper in the case of copper-containing devices), some cramping and pain, expulsion, and pregnancy, including ectopic pregnancy. After IUD removal fertility is usually unaffected. The most serious problems are associated with perforation through the uterine wall or cervix and PID due to various infectious agents. Death has occurred in a few patients using certain devices.

7.4 'HORMONAL' METHODS OF REPRODUCTIVE CONTROL

Interference with endogenous hormones that control mechanisms leading to ovulation, spermatogenesis, fertilization, gamete transport, implantation, and pregnancy recognition and maintenance are all potential contraceptive methods. In the female, hypothalamo–pituitary–ovarian mechanisms involved in gonadotrophin-stimulated preovulatory follicle development and ovulation take place over a few days and are potentially easier to manipulate than the processes of spermatogenesis and epididymal sperm maturation which take place over a much longer period (see Chapter 2, pages 23–30). Suppression of spermatogenesis by negative feedback effects controlling LH and FSH secretion is clearly possible. It has proved difficult, however, to suppress spermatogenesis without suppressing testosterone production and hence libido, potency, and male behaviour. This may not be important for the immunocastration of animals for herd management and/or population management in wild species (see page 146), but it is unacceptable for human male contraception. In the majority of female mammals, both female sexual behaviour and ovulation involve gonadotrophin stimulation of follicle maturation and oestrogen formation. Suppression of both components of ovarian function is appropriate for control of breeding of pet bitches, rodent pests, and strays. In the human female, libido is maintained not *via* ovarian oestrogens but by adrenal androgens; the suppression of factors leading to ovulation do not in general suppress libido, making such methods acceptable.

7.4.1 Artifical steroids

Methods in women

The possibility of exploiting the activity of steroids to block gonadotrophin release and hence inhibit ovulation became a reality 30 years ago with the development of orally active steroids (see Chapter 4, page 62) in the work pioneered by Pincus, Chang and Rock. Such an oral intake, however, exposes the liver to large amounts of steroid, and the steroid to metabolism in the liver. Newer methods employing injection, depot implantation, skin patches, and vaginal rings are now being exploited. Most methods have used a progestogen with or without an oestrogen. Doses of steroid, particularly the oestrogen component, have been progressively reduced because of the association with adverse cardiovascular effects and the possible induction of breast and endometrial cancer.

A wide variety of formulations and methods of administration have been used. Sequential administration of oestrogen alone followed by oestrogen plus progestogen, followed by a break (a regimen which mimicks the sequence of hormonal changes in the normal menstrual cycle) required very high doses of oestrogen to block gonadotrophin release and has been withdrawn from use. Combined oestrogen/progestogen pills may be taken daily for 21 or 22 days followed by a 7 or 6 day break, when a withdrawal bleed occurs. The pills are started either on the first day of a bleed or on day 5 of a cycle. Such pills may also be taken for longer periods without a break (e.g. 63 days, the so-called tricycle regimen) thus giving fewer bleeds. Biphasic or triphasic regimens may be used. These are similar to the previous method except that combined oestrogen/progestogen pills are used in which the dose and/or the proportion of the two steroids varies over the 21 days. Progestogen alone may be taken continuously or administered by depot injection (e.g. Noristerat, Depoprovera) for release over 8 or 12 weeks. Subdermal implants of silastic rubber capsules, containing for example levonorgestrel, release contraceptive levels of steroid for up to 5 years (Figure 7.3). Biodegradable implants lasting 12–18 months, skin patches, and progesterone- or progestogen-medicated vaginal rings or intrauterine devices are also available.

Both oestrogens and progestogens have also been used as postcoital contraceptives (see also page 136). They are capable of interfering with processes of early pregnancy when given shortly before or after coitus.

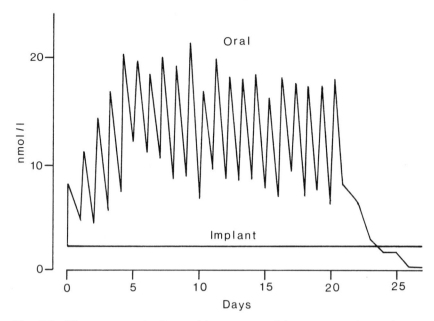

Fig. 7.3 Plasma concentrations of levonorgestrel in women using either an oral regimen or a Norplant implant. (*Data from*: Fathalla, M.F., Rosenfield, A., Indriso, C. (1990) in *FIGO Manual of Human Reproduction*, Vol. 2. Carnforth, Parthenon, p. 75).

Such methods may give unwanted side-effects (e.g. gastrointestinal tract upsets) but may be useful following unprotected/unpremeditated coitus such as rape.

Mode of action of contraceptive steroids

Contraceptives containing both oestrogen and progestogen probably act by depressing pituitary gonadotrophin secretion by negative feedback mechanisms (see Chapter 2, page 20). They may also act by modifying tubal contractions, embryo transport, or endometrial function to adversely affect implantation if ovulation does occur. Progestogen-only preparations, depending on the dose and mode of administration, may act less by preventing ovulation by negative feedback mechanisms and more by either interfering with oestrogenic positive feedback mechanisms, and/or altering cervical mucus to make it hostile and impenetrable by sperm, and/or interfering with sperm transport, capacitation/fertilization and embryo transport, and/or by endometrial desynchronization.

Effectiveness of steroidal contraceptives

Steroidal contraceptive methods are very effective and fully reversible (Table 7.5). Most failures are probably due to user error. Combined oestrogen/progestogen regimens are very tolerant of occasional missed pills, but low-dose progestogen-only regimens are very sensitive to a missed pill, or even to variations in the time of day that pills are taken. Injection/implantation methods are very effective, but there may be a slower return to fertility after withdrawal.

Side-effects of steroidal contraceptives

An objective assessment of the side-effects of steroidal contraceptives is difficult despite their rigorous testing and use for over 30 years. Extrapolation from animal data may be misleading and different preparations have been used by different groups of women over varying

Table 7.3 Side-effects reported by different authors for the same combined oral contraceptive pill (4 mg norethisterone acetate: 0.05 mg ethinyl oestradiol)

Side-effect	Percentage of users experiencing side-effect	
	Lowest	*Highest*
First cycle nausea	1.2	25
Breast discomfort	1.8	13
Weight gain	1.5	54
Menstrual spotting	3.0	17
Breakthrough bleeding	2.1	5.2
Amenorrhoea	0.8	3.6

(*Data from*: Jeffery, J., Kloppr, A.I. (1968) *J. Reprod. Fertil.*, **Suppl. 4**)

Table 7.4 Relative risk of non-fatal myocardial infarction in contraceptive users and others

Group	Relative risk
Controls	1
Oral contraceptive (OC) users	3
Smokers (non-OC users)	5
Hypertensives (non-OC users)	8
Hypertensive smokers using oral contraceptives	170

(*Data from*: Potts, M., Diggory, P. (1983) *Textbook of Contraceptive Practice*, 2nd Edn. Cambridge, University Press, p. 174)

time periods. Even with the same pill side-effects may be reported or important to one group and not another (Table 7.3). It is also very difficult to prove a negative effect. As with any drug, no steroidal contraceptive should be regarded as being without side-effects but it is important to place reports of problems and risks in context with other risks (e.g. travel, smoking, alcohol, see Tables 7.4, 7.6) and with the benefits that may result from taking contraceptive pills.

Useful side-effects
The most useful 'side-effect' for many pill users has been the freedom from worry about pregnancy and hence an increased harmony of many relationships. Some women also have reduced period pain, and relief from ovulation pain (the mittelschmerz) and premenstrual tension. The 'induced' menstrual bleeds are also very regular and may be lighter than normal bleeds.

Annoying side-effects and contraindications
Annoying side effects include breast soreness, weight gain due to fluid retention, nausea, behavioural/mood changes, headaches, blood spotting, and other rarer conditions. Absolute contraindications to pill use include cardiovascular and cerebrovascular disease, reproductive cancers, liver disease, and may include hypertension, diabetes, and older heavy smokers.

Major worries associated with pill use
Failure to begin normal ovulatory cycles after pill withdrawal, so-called post-pill amenorrhoea, may occur in up to 3% of women, depending on the preparation used and the duration of use. Such problems are usually followed by a spontaneous recovery in most cases.

There is no evidence that if a steroidal contraceptive fails and the pill is continued during early pregnancy that congenital abnormalities may be induced in the developing embryo.

The association of cardiovascular problems with pill taking probably relates primarily to the oestrogen-containing preparations. There is evidence that high doses of oestrogen can increase blood coagulability and thromboembolic disorders. If other conditions that may be associated with similar disorders (obesity, age, varicose veins, heavy smoking) are present then oestrogen-containing preparations are contraindicated. Such considerations are clearly illustrated in Table 7.4, which shows the relative single and combined risks of oral contraceptives and other factors of subjects to myocardial infarction.

There are worries that the use of steroidal contraception, particularly those containing oestrogens, may increase the risk of cancer of the breast, cervix, endometrium, or ovary. There is no doubt that factors affecting reproductive status during a lifetime (e.g. ovariectomy, age at first pregnancy, and pattern of child bearing) can alter the risk of cancer of the breast and reproductive tract. The cancer risk for contraceptive users is, however, still unresolved. Combined oral contraceptives do not seem to have any general effect on breast cancer risk (although prolonged early use before a first pregnancy may be harmful), but the data are heterogeneous and may be subject to bias. High doses of progestogen in combined preparations may have a protective effect on the development of benign breast disease. There may be an increased risk of development of endometrial cancer proportional to use, but there may be protection rather than increased risk with oral contraceptives for the development of endometrial or ovarian cancer.

Fig. 7.4 ''Of course, he had to try and go one better and fit a coil''. (*From*: Pyne, K. (1991) *Punch*, Suppl 20–26 November, p. 7).

It is possible that pill users may have an increased risk of catching AIDS. It is suggested that suppression of uterine immunity may make it easier for infection to occur after exposure to HIV.

Methods in men

The use of steroidal contraceptives (progestogens) to block spermatogenesis in men also reduces libido, and replacement with testosterone is associated with possible problems of spermatogenesis maintenance. It may be possible to find a dose of testosterone ester (e.g. testosterone oenanthate) with or without a progestogen, which is acceptable, which will induce a level of oligozoospermia adequate for contraception, and which will also enable full spermatogenic recovery on steroid withdrawal. A more specific target for the inhibition of spermatogenesis is desirable (Figure 7.4). The possible inhibition of spermiogenesis, the post-meiotic production of sperm from spermatids, has been suggested.

7.4.2   Other hormone and hormone-related methods

Breast feeding

Lactation leads to delayed ovulation postpartum (see Chapter 3, page 49). The use of breast feeding rather than bottle feeding has, therefore, been an important component of fertility control in some communities.

Melatonin

Elevated nocturnal concentrations of melatonin interrupt gonadotrophin secretion. The use of nightly doses of melatonin as a contraceptive method is being tested.

GnRH agonists and antagonists

The sustained action of GnRH on the pituitary is to block gonadotrophin release by GnRH receptor down-regulation. With an appropriate dispensing method GnRH or its analogues may, therefore, be useful contraceptive agents. In men, the use of such a regimen to block gonadotrophin secretion could be supplemented by enough testosterone to maintain libido but not to maintain spermatogenesis. This may be possible if GnRH treatment leading to spermatogenic loss is followed by delayed testosterone supplementation. Such a delayed testosterone treatment may be unable to reinstate spermatogenesis.

Inhibin

Inhibin is involved in FSH release inhibition, and with analogue development and/or a suitable dispensing method, may form the basis of a contraceptive technique.

Immunization against gonadotrophins and GnRH

Attempts have been made to block spermatogenesis by immunization against FSH and to immunocastrate animals by immunization against gonadotrophins and GnRH. Immunization to specifically reduce spermatogenesis, as with the use of steroids, presents all the potential problems of disturbing the inter-relationships between FSH, LH, androgens, and spermatogenesis. Immunocastration is often performed in farm animals to remove male behaviour; hence, removal of both spermatogenesis and behavioural aspects of testicular function is often appropiate. Immunization against GnRH rather than against gonadotrophins has proved most successful as such a 'chemical' castration procedure. Another potential use of immunocastration is in the control of pest species, when such sterile males could be produced and released into the wild. To be effective, however, such males would need to compete for and mate with females (e.g. producing pseudopregnancy in rodents, see Chapter 3, page 46) but not produce sperm. Hence a selective spermatogenic block will be required.

Gossypol

This yellow pigment from chinese cotton-seed causes infertility in men. In trials, over 99% of men taking gossypol showed a fall in sperm count to levels believed to be incompatible with fertility. The effects were reversible. A significant side-effect, however, is a fall in blood potassium, which may be due to an inhibition of corticosteroid synthesis. It may be possible to find a component of gossypol with antifertility but not potassium-lowering effects.

Once a month pills

Compounds that interrupt progesterone synthesis (e.g. epostane, see page 65) and antiprogestogens (e.g. RU486, see page 65) will interrupt the cycle if given during the luteal phase, and inhibit ovulation if given during the follicular phase. They are therefore presumptive contraceptive agents. RU486 has, however, been shown to have a low contraceptive efficacy when given as either a single high dose or repeated doses.

7.5 POTENTIAL CONTRACEPTIVE METHODS AFFECTING SPERM MATURATION AND FERTILIZATION

Spermatozoa undergo extensive maturation in the epididymis and in the female tract (see Chapter 2, pages 29 and 35). Only capacitated spermatozoa can undergo the acrosome reaction leading to fertilization which follows an initial sperm–zona pellucida recognition involving specific sperm and zona binding sites. Hence, knowledge is accumulating for specific loci for potential contraceptive methods.

Sperm may be blocked by the development of specific antisperm antibodies or other antisperm agents. Epididymal function and sperm maturation therein is controlled by androgens. Androgens reach the epididymis not only *via* the peripheral blood but also attached to androgen-binding protein *via* the seminiferous tubules. This suggests that the epididymis may have a high requirement for androgen compared with other androgen-dependent systems. Attempts have been made to lower sperm maturation but retain libido in men using anti-

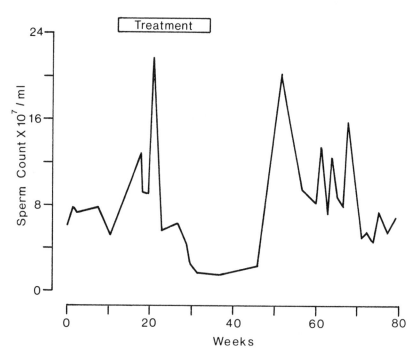

Fig. 7.5 The effect of an antiandrogen (cyproterone acetate) on sperm production in man. (*Data from*: Morse, H.C., Leach, D.R., Rowley, M.J., Heller, C.G. (1973) *J. Reprod. Fertil.*, **32**, 368).

androgens (Figure 7.5). Such a method may also be used to produce 'sterile males' for pest control.

Capacitation involves changes in surface components of the sperm head which may be manipulable. The development of antibodies against the specific protein recognition sites on the sperm and zona pellucida prior to fertilization may provide a contraceptive method with high selectivity. Zona pellucida antibodies have contraceptive activity, inducing infertility *in vivo* in the marmoset and inhibiting human fertilization *in vitro*. The former also caused a depletion of the primordial follicular pool, suggesting that more selectivity is required.

7.6 COMPARATIVE EFFECTIVENESS OF EXISTING CONTRACEPTIVE METHODS

Sterilization correctly performed will yield 100% effective contraception. The effectiveness of other methods depends on both the inherent efficiency of the method and the motivation and experience of the user. Table 7.5 shows potential typical and lower levels for efficiency of the well-tried contraceptive methods based on failure rates.

7.7 CONTRACEPTIVE RISKS COMPARED TO OTHER RISKS

Some methods of contraception, such as the use of steroids, are considered by some to be of high risk, and reports of problems associated

Table 7.5 Relative effectiveness of different contraceptive methods (first year failure rates)

Methods	Failure rates (%)	
	Lowest	*Typical users*
None	90	90
Coitus interruptus	16	23
Sponge with spermicides	9–11	10–20
Periodic abstinence	2–20	24
Condom (♂)	2	10
Diaphragm/Caps (♀)	2	13–19
IUD	1.5	3–5
Steroidal methods		
Oestrogen + Progestogen pills	0.5	2
Progestogen only pills	1	2.5
Injectable progestogens	0.25	0.25
Sterilization	0.1	0.4

(*Modified from*: Fathalla, M.F., Rosenfield, A., Indriso, C., eds. (1990) *FIGO Manual of Human Reproduction* Vol. 2, Parthenon, p. 23)

Table 7.6 Life expectancy based on a single risk of death*

Risk	Life expectancy (years)
Smoking 40 cigarettes per day	100
Riding a motorcycle 10 h/week	300
Drinking a bottle of wine per day	1,300
Driving a car 10 h/week	3,500
Power-boating once per month	6,000
Having a baby every year	10,000
Playing football twice a week	25,000
Staying at home watching television 200 h/year	50,000
Using oral contraceptive (non-smoker)**	63,000
Travelling by train 100 h/year	200,000
Using an IUD	200,000
Being struck by lightning	10,000,000
Being hit by a falling plane	50,000,000

*US/UK data
**Considers only cardiovascular risks
(*From*: Potts, M., Diggory, P. (1983) *Textbook of Contraceptive Practice* 2nd Edn Cambridge, University Press, p. 391)

with their use receive wide coverage in the press. Though not wishing to underplay any increased risks, it is probably important to put such risks in perspective relative to other every-day risks. By the year 2000, for example, more men in the UK will have died from smoking-induced cancer than from two world wars. For every 10 cigarettes smoked by a pregnant woman her baby will be 0.1 g underweight at birth. If expectations of life based on a single risk are compared it can be seen (Table 7.6) that contraceptive risks compare favourably with those of some other activities.

CHAPTER 8

Social and ethical considerations

Biology in the 1990s is in many ways at a comparable point to that of physics in the 1930s, when man could split the atom to release huge amounts of energy, exploitable both for peaceful means and for warfare. It is now clear that the full implications of both were not clearly understood by many politicians or by the majority of the public. A desire to end World War II quickly led to the use of atomic bombs. Subsequently horror over their effects led to both their use as deterrents and cries for their abandonment. Peaceful uses of atomic energy for the production of isotopes used in medicine and agriculture have been exploited, but initial enthusiasms for atomic power stations have been muted after accidents at Three Mile Island and Chernobyl, and the realization that dangerous wastes have to be retained. Similar dilemas face biologists and society today. The possibilities for exploiting new techniques in the field of reproductive manipulation and elsewhere are incalculable and the potential benefits to mankind enormous. The implications of potential advances must be, however, fully spelt out so that society can decide how far it wishes to go in the 'manipulation' of either human or animal life. It is also important that the quest for knowledge in this area of science is not taken as implying any desire for misuse, and that the possibility of misuse does not delay the obtaining and appropriate application of valuable knowledge. Informed education is very important. Reproduction and genetic engineering developments can give rise to sensational press reports. It is very important that the real facts and the balanced issues are discussed.

8.1 THE GENERAL ISSUES

It is clear that in the whole area of intervention in reproduction there are strongly held views taken on religious and moral grounds and in the support of individual rights. The ethical and moral high ground is frequently claimed by advocates with totally opposed views – either for or against contraception or for or against abortion. If one tries to find the ethical norm, how does one decide? Is there an effective mechanism for formulating public policy in the area of reproductive manipulation? Is a primitive society practicing cannibalism and infanticide any worse than sophisticated but modern countries allowing wars and famines? Is the choice of controlling population by therapeutic abortion, preimplantation embryo removal, or prefertilization block of gametes, only one of degree? When is a human being formed – at fertilization, implantation, when a nervous response takes place or when a human-like form has differentiated? Do we do nothing to interfere with life, or do it as early as possible, or do it at almost anytime for the right reasons, or give people the choice? If we decide, after much heart searching, that very gross abnormalities should not be aborted but that an anencephalic newborn should not be sustained, do we condone research that may lead to an intrauterine cure? How can human reproductive technology advance without undermining deeply held beliefs? Can assurance be given that advances will be positive? For the formulation of policies we need to be aware of the moral sensitivities of society, and need to explain the potential benefits and risks of planned procedures within a widespread public debate, before guidelines are devised.

Attitudes to intervention in human reproduction may be very different to attitudes taken towards manipulation of animal reproduction. Man has a moral obligation to animals: to conserve species and habitats and to prevent ill treatment of either individuals or groups of animals. A quite proper interest in animal rights and welfare would not exclude, however, the development of clones of a domestic animal, whereas similar work on the human would be abhorrent. Some would advocate no interference with or exploitation of animals, whereas others would see the manipulation of domestic farm animals and the control of pests as desirable, but be against interfering with or exploiting wild animals. Attitudes have, however, changed towards wild animals from the 'exploitation versus preservation' arguments to a balanced conservation stance in which the progressive perilous interaction of many species with man may entail elements of management, including captive propagation, and the setting up of resource banks of embryos and sperm for their continued existence.

Some may advocate degrees of non-use of domestic animals for food, fibre, or work, but the use of domestic farm animals by man is accepted in many countries, without questioning the ethics. Increased interest in animal welfare has rightly questioned the use of intensive inadequate conditions for maintaining animals. Efficiency is, however, not incompatible with compassion, and reproductive manipulation can be part of good management. There will, however, probably be an increasing emphasis on the use of non-invasive methods, which are seen to be more natural, and which avoid the possibility of drugs and their metabolites entering the food chain. An additional legal dilema has arisen regarding the production of genetically modified animals. Should such procedures be regarded as inventions and hence patentable or should they be regarded as discoveries and remain the property of all?

8.2 PROBLEM AREAS IN THE MANIPULATION OF REPRODUCTION

When it comes to a consideration of the ethical and moral views of reproductive manipulation no view is right or wrong and a whole range of views may prevail. Considerations may, however, be at two levels: (i) as they affect the individual animal, couple, donor, or surrogate, or as they involve an individual procedure or method (abortion, sterilization, use of drugs, 'natural methods); and (ii) as they affect group views and possible generalizations relating to human rights, politico-religious considerations, guidelines, and laws.

Such considerations applied to the manipulation of human reproduction involve the need for consensus and for legal definitions of when life begins and when life can be manipulated or terminated. Three areas where intervention is possible warrent consideration: first, the prevention of birth, including contraception, sterilization, abortion, or selection for birth after diagnosis of congenital disease or multiple fetation; second, assistance with conception, including hormone therapy, artificial insemination, and IVF and its variants, including surrogacy; third, manipulation of embryos, including sex selection, congenital disease selection, freezing, genetic manipulation, and the use of embryos for research.

8.2.1 'Rights' of the fetus

What are the rights of an unborn fetus? Some may feel that the mother has the rights and the choices, and that 'abortion on demand' should be legal. At the other extreme some feel that abortion, even on the grounds of a potential very serious handicap, should not be permitted. All strands of belief within this wide range are held. Many may agree that a fetus can be aborted if it has Down's syndrome, but would oppose the selective reduction of multiple fetuses produced as a result of IVF or GIFT. Society may take a strong line on abortion but only encourage and not restrain mothers from potentially harming their babies due to smoking or using alcohol or drugs. Clearly the rights of the unborn are often viewed differently to those of the born baby. Most, if not all, of those who condone abortion under specific circumstances would never condone infanticide for any reason.

8.2.2 *In vitro* fertilization

Any treatment for infertility is as ethical as any other medical treatment if the rights to have a family are placed on equal footing with the rights to have a disease controlled. *In vitro* fertilization entails, however, manipulation outside the body of both fertilization and early embryo development, which some may regard as inappropriate intervention. There is also the dilema of what happens to spare embryos. Should they be stored and who 'owns' them? When embryos are replaced in the mother, how many should be replaced? Improving the chances for a single offspring by replacing more than one embryo clearly increases the risk of multiple births. Can differential abortion be justified under these circumstances?

8.2.3 Sperm, egg and embryo donation

Artificial insemination by the husband has potentially few ethico-legal problems. Does AIH, however, consummate a marriage and does a child born after a father's death and conceived with his frozen sperm have equal inheritance and other rights? Artificial insemination by a donor may raise additional considerations. Can it constitute adultery if performed without the husband's consent and could the child be illegitimate? Should a child born following a donor insemination have the right to know the identity of the father? Under such circumstances are men less willing to become sperm donors? In countries where this is the law the number of births from donor inseminations has fallen. Finally, should insemination of single women be permitted? If natural insemination of single women is legal, donor insemination, though not perhaps favoured by some, cannot surely be illegal. When the possibilities of oocyte and embryo donation are added to sperm donation the situation can become progressively more complex (Table 8.1). IVF or related techniques may allow a woman to give birth to a baby derived from her oocyte and her husband's sperm, her oocyte and another man's sperm, another woman's oocyte (could be from a close relative, e.g. sister) and her husband's sperm, or totally 'unrelated' gametes from another man and another woman. As the 'wife' gives birth to the baby she is regarded as the mother. Does the position change, however, if a surrogate uterus is used? Can such a woman be made to give up the baby? Does it affect the issue if money changes hands? Some feel particularly uneasy about the latter. If sperm and/or oocytes, and/or embryos are (or can be) frozen, additional complexities arise. Embryos in deep-freezes could be potential recipients of inheritances and a mother could give birth to an identical twin sister who has been frozen since the mother was transferred to their mother, when the twin sister was a spare embryo and frozen.

Table 8.1 Some alternative methods for reproduction

Sperm source	Oocyte source	Fertilization site	Pregnancy site	Assistance method involved
Husband	Wife	Wife	Wife	None, AIH or GIFT
Another man	Wife	Wife	Wife	AID
Husband	Wife	In vitro	Wife	IVF, ZIFT
Another man	Wife	In vitro	Wife	IVF, ZIFT + donor sperm
Husband	Another woman	In vitro	Wife	IVF, ZIFT + donor oocyte
Another man	Another woman	In vitro	Wife	IVF, ZIFT + donor gametes or embryos
Husband	Another woman	Another woman	Wife	AIH + donor woman
Another man	Another woman	Another woman	Wife	AID + donor woman
Husband	Wife	Wife	Another woman	Surrogate mother
Another man	Wife	Wife	Another woman	Surrogate mother
Husband	Wife	In vitro	Another woman	Surrogate mother
Another man	Wife	In vitro	Another woman	Surrogate mother
Husband	Another woman	In vitro	Another woman	Surrogate mother
Another man	Another woman	In vitro	Another woman	Surrogate mother
Husband	Another woman	Another woman	Another woman	Surrogate mother
Another man	Another woman	Another woman	Another woman	Surrogate mother

(*From*: Wallach, E.E. (1990) in *The Johns Hopkins Handbook of in vitro Fertilization and Assisted Reproduction Techniques* (ed M.D. Damewood) Boston, Little Brown, pp. 151–62)

8.2.4   Sex determination and embryo manipulation

Sex determination and selection of sperm or embryos linked with AI and/or related procedures enables the sex of a baby to be chosen. Is it acceptable to offer such a choice and what impact might it have? In some societies a boy may be socially more valuable than a girl. Choice of sex could therefore upset the sex ratio. Embryo manipulation for diagnosis of genetic abnormality and for research is considered acceptable by many, under controlled conditions. Certain possible manipulations such as cloning of humans are, however, regarded with foreboding and abhorrence.

8.2.5   Extension of existing techniques

The possibilities of ovarian and testicular culture, linked to IVF and related methods, followed by the use of artificial uteri and placentae, may have to be faced in the future. Many who regard existing technologies as appropriate for animal genetic improvement, conservation and reproductive efficiency, and with reservations (e.g. cloning) as appropriate for human use, may have increasing reservations as the deviations from the 'normal' become more and more severe.

8.3   LEGAL RESTRAINTS AND GUIDELINES

If reservations exist with respect to the use of contraception, abortion, and surrogacy how should controls be established? The laws of many countries are still governed by traditions of Judaeo-Christian morality, even where a majority of citizens are no longer active proponents of the faiths. Under such circumstances do anti-abortion or anti-contraception laws make for less abortion and less contraception, or perhaps for more law breaking and more medical risk? More liberal laws, allowing people to do things that may seem reasonable to a middle majority, do not imply that people have to do things that they find ethically or morally wrong. Laws should try to strike a cautious balance reflecting a majority view.

Most people would agree that in a developing and complex field of activity such as reproductive manipulation 'the law' should take time to be formalized. It took 9 years from the birth of Louise Brown in 1978 for the first legislation to regulate IVF procedures to be formulated, in Norway. The situation has now been extensively considered in many countries, and guidelines, watch-dog authorities, and licensing procedures have been set up. One problem has been in appropriately defining established procedures, such as IVF, so that variants on it or newer procedures, such as GIFT, can be included or added without undue delay. As even many biologists may disagree on the precise difference between an embryo and a fetus the problems for legislators may be high. Effective regulations, therefore, will probably depend on a balance in legislation that, on the one hand avoids abuse and gives protection to embryos, women and clinical and research workers, and on the other, places confidence in and gives opportunities to the latter for advances to take place.

There are trends in many countries for the acceptance of AIH and AID, for recognizing that a child from AID with the husband's permission is his legitimate offspring, and for accepting IVF, if performed for married couples and if gamete donors are not used. There is less general acceptance for surrogate motherhood and ovum donation, particularly if money changes hands. There seems to be no consensus, for or against, research on human embryos.

8.4 ATTITUDES OF THE USER

Despite certain moral, religious and political pressures for or against family limitation, autonomous reproduction has been regarded as a basic human right along with the right to health, food, and shelter. Is there, therefore, an equal right to contraception or conception assistance, or are these the responsibility of the individual? Can the world afford, however, not to undertake the former and can the world justify spending on the latter compared to the priority of basic health care? How logical is it to accept all the implications of an obstetrical service yet pay less regard to those unable to take part in the service through no fault of their own? In an overcrowded world the plight of the childless may be overlooked.

Acceptance of contraception depends on a whole array of factors and issues. It is essential that knowledge of the possibilities for family control are available to all and that a method is perceived as being effective, convenient, and safe. Overall approval of the regulation of family size on religious or ethical grounds is linked to individual ideas of the right family size, the need for sons, attitudes to sex and sex roles, partner approval, and personality variations, such as a desire to conform or not conform to the norm, or to be essentially careful or fatalistic. Personality factors also play an important part in the choice of contraceptive method, and some of the adverse reactions and reasons to discontinue a method reflect personality type. A proportion of women react adversely to steroidal contraceptives with depressive mood changes, loss of sexual desire, or both.

Infertile couples often have repeated reminders of their position from potential grandparents and the rest of society, which in many countries is still represented by family units. In addition, a woman or a couple may feel a real need for a baby. This may begin to rule their lives, and lead to blame on one side or the other and eventually to marriage problems or a marriage crisis. Are these valid reasons, however, for society to give priorities in this area? If available to them, some couples, though initially motivated to attend an assisted conception programme, monitoring, drug therapy, collection of oocytes for IVF, etc., may come to find that the effect on their lives and bodies is too high a personal price to pay, and the monetary cost for perhaps limited potential success may also seem too high. There is the worry that the high degree of motivation for a baby, at any price by some couples, may be exploited. Are methods like IVF and GIFT always performed after an appropriate preliminary investigation to ascertain the appropriateness of treatment? Clearly this is the case at most centres, but all may not be so good. If there is a conflict for funding for fertility assistance for women, should efforts be concentrated, for example, on the prevention of blocked tubes rather than treatment of the condition?

8.5   IMPACT OF AND DEMAND FOR CONTRACEPTION

The extended development and availability of sterilization, IUDs, oral and injectable steroids, condoms, abortion, and newer techniques has had a profound impact. By 1980, contraceptive users in less developed countries outnumbered those in developed countries by 3:1: this represents a 10-fold increase in total use over two decades. There was also a change from male user to female user to make female:male responsibility as high as 3:1.

In the 30 years generation from 1980 to 2010, the number of women (already born) who will be of reproductive age may have doubled to around 800 million. If trends continue there may be 450 million couples using family planning by 2010 (Figure 8.1). A qualitative change is also expected as urbanization and education are increased (Table 8.2). In this regard, the importance of sex education can be emphasized. Improved sex education in schools can reduce a rising number of unwanted pregnancies. Young people have strong sex drives and intercourse is related to sex drive not planning. Information on the biology of sex must be supplemented with information on the nature of sexuality and human relationships.

The effectiveness and impact of contraception would clearly improve if there was an increased adoption of contraception by non-users and a decreased rejection by users. Such changed attitudes should be voluntary, but have also been achieved by political intervention. In China dramatic changes in population policy have occurred and been shown to be effective. Initially intervention in population policy increased population in the period from 1949–1970. Subsequently reductions

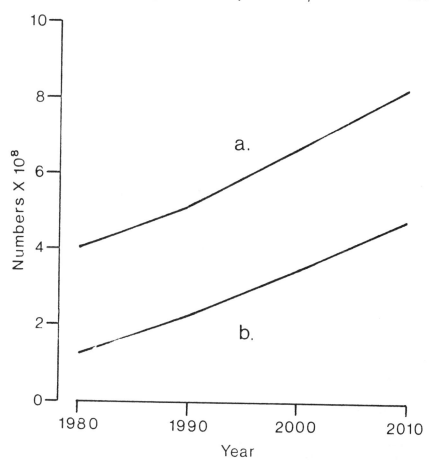

Fig. 8.1 A projection of possible numbers (a) of women of reproductive age and (b) family planning users. Data from less developed countries, excluding the People's Republic of China. (*Data from*: Gillespie, D.G., Cross, H.E., Crowley, J.G., Radloff, S.R. (1989) in *Demographic and Programmatic Consequences of Contraceptive Innovations* (eds S.J. Segal, O.A. Tsui, S.M. Rogers) NY, Plenum Press, p. 268).

occurred due to advocacy of late marriage, but primarily by the one child per family policy backed by both incentives and penalties. Increased voluntary use of contraceptives can be encouraged by the ready availability of methods perceived by the potential user to be both effective and acceptable. Such methods have a low failure rate and low incidence of side-effects and must be variable in design and use, to meet all needs. Some methods, though effective, may be inconvenient or unacceptable involving, for example, those that require multi-pill taking, are coitus related (condom), or which require the woman to touch her genitalia. Linking contraceptive methods and family plan-

Table 8.2 Population, life expectancy, urbanization and family planning projections 1980–2010*

Parameter	1980	1990	2000	2010
Population ($\times 10^6$)	2300	2900	3600	4300
Life expectancy (years)	54	57	61	64
Urbanization (%)	33	38	44	51
Family planning users ($\times 10^6$)	129	227	349	476

*Data for developing countries excluding People's Republic of China
(*Data adapted from*: Gillespie, D.G., Cross, H.E., Crowley, J.G. and Radloff, S.R. (1989) in *Demographic and Programmatic Consequences of Contraceptive Innovations* (eds S.J. Segal, O.A. Tsui, S.M. Rogers) NY, Plenum Press, p. 269)

ning to preventative medicine and screening facilities, to changes in social and religious attitudes, to attitudes towards sexual practices, and to the impact of such factors as earlier puberty and AIDS may all encourage use. Above all there is the problem of supply, and of fulfilling the potential demand. Such needs may take at least double the current costs. In the 1980s costs were met by governments, donor governments, and by commercial sources, and many new products were developed. Despite the continued need for the development of new methods, support is declining. Government supported research of all kinds is declining in many countries and large companies have withdrawn support for contraceptive research, due to legal liability claims and the costs and delays of putting a new drug through toxicity trials and licensing procedures. Hence the funding to supply existing methods and the prospects for new breakthroughs seem difficult. The situation perhaps calls for more action by the United Nations. Originated to maintain world peace, it has agencies for world health and food and welfare. Helping to provide appropriate contraception may aid its activities in each of these areas.

8.6 COSTS OF FERTILITY INDUCTION

Costs related to fertility induction compared to world family planning may seem less formidable and less urgent. The application of schemes for farm animals involving reproductive manipulation will always be dependent upon their cost-efficiency in terms of potential genetic improvement, efficiency, or convenience. Application of methods for wild animal conservation may gain both public and private additional financial support as the public face the prospect of well-loved species facing extinction. The cost of financing human conception assistance is becoming progressively regarded as something that warrents public research support but not public health service support for the individual user. Failure to conceive does not, in the view of many, have as high a priority as health care or disease prevention, yet obstetrical care and child care is never questioned.

CHAPTER 9

Conclusions and future prospects

Reproductive biology faces an array of challenges, including the development of efficient and acceptable methods of family planning and aleviating infertility in man; improving fertility, production and breeding potential of domestic species; and conserving and managing various wild, feral, and endangered species. The problem of population increase in particular impinges on and exacerbates many other problems. There is some evidence that in developed countries economic and social progress has lowered population growth and that population problems could be solved by rapid economic growth. It is probably true that economic and social progress helps slow population growth, but rapid population growth also slows economic development. Population increase, if sustained, can overtake any economic and social progress generated: the increased need for land and water, the over-exploitation of resources and the over-loading of land leads to pollution, deforestation, loss of grazing, soil erosion, and extinction of species. The long-term solution may lie rather in the reinstatement of more balanced communities in which population, food supply, agricultural systems, and the environment are more in equilibrium. There is a need to look at economic activity, environmental conservation, and population growth as linked issues.

The challenge facing reproductive biologists is one of a greater and more specific knowledge of all aspects of the ecology, physiology, and endocrinology of reproduction in a wider array of species, enabling an ever-expanding knowledge base to be applied for the management and control of reproduction.

The reproductive processes in mammals are very diverse and only a very few species have been studied in detail. Extrapolation of findings across species is very dangerous, but understanding of similarities and differences between different species can greatly advance individual understanding and allow generalizations and general solutions to be formulated. Apparently unusual features of reproduction such as the fact that the plains viscacha ovulates between 300 and 700 eggs at each oestrus but that only two young are carried to term, or that intrafollicular fertilization occurs in two species of Malagasy hedgehogs, are not only important for understanding the specifics of one species, but will often give important clues as to how certain features of reproduction may be naturally or artificially manipulated in others. Techniques such as AI, IVF, and embryo transfer, developed in a non-endangered model species, such as the common ferret and domestic cat, can be applied to related wild and endangered species such as the black-footed ferret and tiger, respectively.

Much is known of the gametogenic and hormonal functions of the gonads in several species, and of their hormonal control. Much of the latter information is, however, based on *in vivo* or *in vitro* model systems which illustrate the possible mechanisms present. The detailed, integrated, sequential relationships involved in the control of gonadal function, particularly at the autocrine and paracrine levels, await clarification.

The internal reproductive control mechanisms are influenced, to a different degree in different species, by various social and behavioural factors interacting with an array of environmental factors, such as photoperiod, climate, nutrition, and stress. The primary link between these external factors and the internal control system of the gonads is the extrahypothalamic–hypothalamic–pituitary axis, which exerts its effect through changes in the pulse generator and the episodic output of GnRH. The details of the various external–internal links and the role of specific neurotransmitters, opiate peptides, feedback signals, and endogenous rhythms await further clarification and may vary greatly in different situations. Whereas the social/pheromonal ('ram effect') signal influencing the onset of ovarian activity in the ewe leads to a very rapid increase in LH episodic output, suggesting a very direct stimulation to, or inhibition removal from the pulse generator, the photoperiodic seasonal effect, mediated by melatonin, is much slower and probably represents a resetting of an endogenous rhythm. The primary action of melatonin may be on cells of the pars tuberalis of the pituitary, with an as yet unknown link to the 'rhythmic centre' in the hypothalamus. Whether different nutritional influences on reproduction operate *via* the pulse generator mechanism and/or *via* components of the growth hormone/IGF-I system and/or elsewhere awaits further study.

In the context of gonadal function the interaction between the extragonadal (gonadotrophin) control system and intragonadal mechanisms is becoming clearer.

In the male, knowledge of the interaction of an FSH-stimulated Sertoli cell-dependent mechanism with an LH-stimulated Leydig cell-dependent mechanism involving testosterone for the control of spermatogenesis, is being extended to the understanding that interactions between peritubular cells, Leydig cells, Sertoli cells and spermatogenic elements are involved in the control of spermatogenesis. An understanding of the mechanisms involved in the control of spermiogenesis and the spermatogenic cycle may offer specific loci for manipulation.

In the female, the control of the early stages of folliculogenesis probably primarily involves intraovarian mechanisms. Waves of development of antral follicles and preovulatory follicular recruitment (and dominance where appropriate) involve pituitary gonadotrophins modulated and mediated by a range of steroidal, peptide, and other intraovarian mechanisms. Species differences are particularly noticeable in the control of ovarian function, where differential follicular recruitment mechanisms lead to single, a few, or many preovulatory follicles, and where ovulation may be caused either *via* an oestrogen-induced or coitus-induced preovulatory LH surge. Corpus luteum function also varies greatly between species. Wide variation is seen in the components and stimulus of the luteotrophic complex involved in luteal maintenance, in the mechanisms involved in luteolysis in an infertile cycle, and in the methods of conceptus signalling for corpus luteal maintenance in early pregnancy. Interferon-like substances involved in luteal maintenance in some species may have a more general role in early pregnancy in others. Later in pregnancy progesterone levels are maintained by either a continuance of luteal function and/or the production of progesterone by the placenta. This difference in pregnancy maintenance is partly mirrored in the mechanisms controlling the onset of parturition where, at least in some species, a fetal-initiated rise in corticosteroids leads to a progesterone fall, an oestrogen rise, and a rise in prostaglandin; in other species such mechanisms may be less important.

Limitations in reproductive activity in mammals include periods of natural quiescence, such as between ovulatory episodes in non-pregnant menstrual or oestrous cycles, in the period prior to puberty, during pregnancy and into senescence; and periods of externally influenced quiescence, involving suckling, seasonal, environmental or social intervention; specific congenital, metabolic or disease related syndromes; or voluntary abstinence or intervention by contraceptive means. An understanding of the mechanisms involved in the onset of quiescence, or the 'recovery' period following quiescence, can give valuable insights into possible methods of artificially controlling reproduction. It is of interest to know whether the mechanisms of reproductive suppression before puberty, during seasonal inactivity, during post-partum inactivity, and during the period approaching senescence are related; and whether reinitiation of reproductive activity after such conditions involves similar mechanisms, such as an opiate involvement. Although clear differences are being shown, it is not known how such mechanisms react with the final common pathway of action, involving the pulse generator and the episodic output of GnRH, if such a single pathway exists. A clear understanding of this complex integrated network will reveal more precise loci for manipulation.

Possible sites for reproductive manipulation involve every facet of the reproductive system, its control mechanisms in the male and female, and their interaction. Modifications can be made to social, behavioural, ecological, and environmental factors affecting reproduction and the mechanisms that mediate them *via* all aspects of the hypothalamo–pituitary–gonadal axis. Factors affecting gametogenesis and the hormonal functions of the gonads (behaviour, accessory glands), sperm transport and maturation within the male tract, sperm transfer to the female, sperm transport and maturation within the female tract, fertilization and pregnancy can all be manipulated. Various methods can also be used to manipulate gametes and early embryos *in vitro*, involving AI, IVF, freezing, and sexing, and future developments may include the culture of embryos, artificial or cultured endometrium, and perhaps even maintenance of spermatogenesis and folliculogenesis in culture. Although there has been a move towards the use of natural non-invasive methods for both fertility induction (involving social, behavioural, and pheromonal or genetic methods) and for human family planning, much manipulation of reproduction may continue to be through direct or indirect alterations of the hormonal control systems.

Naturally occurring hormones or their natural or synthetic analogues (which may be administered by mouth or have a longer half-life and more sustained action) have been used. Substances that stimulate or inhibit hormonal synthesis, secretion, blood concentration or action, and active and passive immunization against specific substances have also been employed.

The actions of hormones and hormone analogues depend on their mode of administration, dosage, and metabolism. Highly active antagonists and agonists of naturally occurring substances (e.g. GnRH) and orally active analogues and derivates (e.g. contraceptive steroids) have been prepared and various active analogues of other substances (e.g. naphthalene derivatives which mimic melatonin action) are being studied, along with novel methods of administration by patches, minipumps, and slow release and differential release materials, such as soluble glasses.

Methods of intervention in reproduction, for whatever purpose, should be effective, safe, without side-effects, acceptable and, as appropriate, reversible. To achieve these criteria, methods are being sought that are highly specific.

The understanding of normal reproduction, the diagnosis of abnormality and the monitoring of the effects and effectiveness of manipulations require reliable and practical methods for the assessment of reproductive function. Methods for the assessment of gonadal function have employed either visible signs of changes in hormonal activity and action (e.g. oestrous behaviour, vaginal mucus, basal body temperature, the onset of menstruation), direct measurements of hormones in body fluids, or even more direct observations of the gonads (e.g. testicular measurement and semen analysis, ovarian palpation, laparoscopy, and ultrasound). The ready availability of semen has led to its extensive use for diagnosis in the male. There is evidence, for example, that the sperm count of western men has fallen by about 50% between 1940 and 1990. This may be caused by a build up of toxins in the environment disrupting testicular function. Ejaculated sperm, however, are epididymal sperm and are far removed in time from the spermatogenic stages. A more direct parameter for spermatogenesis might be the measurement of factors in blood that are linked with the spermatogenic cycle. Classical semen parameters such as volume, live sperm concentration, morphology, and motility may have a poor correlation with fertility. More refined motility evaluation, tests for sperm integrity, and *in vitro* tests for fertilizing capacity, involving hyperactivity testing and the use of homologous and heterologous oocytes and zonae pellucidae, are offering advances particularly when used with IVF, AI, and related procedures.

A major need in the female is a direct practical method for assessing ovulation. Ultrasound, though valuable, has a limited wide practical application, and other methods used, based on oestrous behaviour, cervical mucus, basal body temperature, and steroid determinations, are related to preovulatory folliculogenesis or corpus luteum development, and are not assessments of ovulation *per se*.

Despite the above reservations, hormonal measurements have proved particularly valuable for assessing reproductive function. Measurements are, however, greatly affected by the episodic nature of hormone release, the rate of hormone metabolism and the nature of the sample studied (e.g. blood, urine, saliva). Sequential measurements, studies on more than one component of a system (e.g. the study of both pituitary LH and testicular androgen) and the use of dynamic tests, have all aided the interpretation of assays. The development of radioimmunoassay revolutionized hormone measurement and the use of non-radioactive labels and of body fluids (urine, saliva, faeces) other than blood is leading to the development of a wide range of methods usable in the field, on the farm, and in the home. Such developments herald the possibility of miniaturized biosensors giving continuous readings as required. Assays based on immunological reactions do not, however, necessarily correlate with biological activity, and such assays require validating or cross-checking with receptor assays employing specific binding proteins, or other biological methods.

Accurate and early pregnancy diagnosis is important in man and animals. The best methods will probably depend on an early, either direct observation of the embryo, by ultrasound or some other non-invasive method, or indirect indications of the presence of the embryo by the measurement of a substance unique to pregnancy such as a conceptus signalling substance (e.g. chorionic gonadotrophin, trophoblast protein), rather than changes in a non-unique substance, such as progesterone, which may reflect non-pregnant luteal function. Early diagnosis of pregnancy loss is also important: a method suitable for the early indication of pregnancy may need to be combined with a method to determine continued pregnancy.

Fetal abnormalities can be diagnosed by amniocentesis and chorionic villus sampling or by sampling embryos *in vitro*, combined with analytical methods such as karyotyping or genetic probes. The sexing of sperm and embryos has a wide range of potential applications in man and animals.

Methods developed for inducing fertility or conception assistance in various species can have a number of aims. The objective may be to induce a normal number of gametes in an animal or human in either a natural or an abnormal reproductively quiescent condition, to increase the number of gametes to an optimum for the species (e.g. to obtain twins or triplets rather than singles), to increase the number of gametes (or embryos) to supraoptimal for use in IVF, embryo transfer and related programmes, to alter the timing or synchronization of ovulation within or between individuals for use in IVF and batch artificial insemination programmes, to increase conception rate or reduce embryo loss, to achieve hormone replacement without fertility in human primary hypogonadism and the menopause, or for the preparation of 'teaser' animals.

Specific methods have employed either indirect genetic or environmental stimulation or direct hormonal replacement or manipulation of components of the extrahypothalamic–hypothalamic–pituitary–gonadal axis; or the manipulation of gametes or embryos, including variants on AI, IVF, embryo transfer, freezing, and genetic or other manipulations *in vitro*.

Such methods often lack precision, efficiency, and specificity. The development of more reliable hormonal manipulations for the induction of multiple preovulatory follicles, for example, depends on a clearer understanding of the extra- and intraovarian factors involved in pre-ovulatory follicular recruitment and dominance in different species, and the ability to induce or exaggerate such events by the use of tailor-made recombinant gonadotrophin molecules, so that a correct sequence of stimulation can be given over time. The need to stimulate an ovary containing developing follicles capable of response is also important. The induction of temporary pharmacological hypophysectomy by GnRH analogues shows promise and offers the possiblity not only of obtaining ovaries at a unified appropriate starting state for stimulation, but also of eliminating interference during treatment by endogenous gonadotrophins or feedback mechanisms leading to single follicular dominance.

An extension of the use of *in vitro* techniques can be expected, with refinements in methods for *in vitro* fertilization and early embryo development for an extended range of species, and for the freezing of oocytes as well as embryos and sperm. Techniques will be improved for the transfer and replacement of sperm, oocytes and embryos into individuals of the same and related species. The latter may help endangered species to be saved from extinction. A wide array of sexing, transgenic, and cloning techniques will be possible. Further advances may involve *in vitro* culture of testes and ovaries (and the direct *in vitro* production of oocytes, sperm, and embryos), and the development of natural or artificial endometria, allowing the potential culture of embryos to more advanced stages.

Artificial insemination has led to important advances in animal breeding, particularly in cattle, when linked to the use of oestrous synchronization techniques. Embryo transfer may prove even more important for further developments in sex selection, IVF, and the cryopreservation and manipulation of embryos. The extension of such techniques may involve the use of fewer and more specialized animals for breeding, with special sperm producers, oocyte producers, and incubation females producing all the production stock. Other developments in farm animals may include reducing the time to puberty, shortening the interval between the birth of the offpsring and remating, and the removal of seasonality in species such as sheep and goats.

The ability to permanently or temporarily prevent, block, or defer aspects of reproductive function has a wide application, not only for human population control and family planning, but also for use in working animals (e.g. horses, dogs, camels), in animals used for sport and as pets, and in managing both endangered and pest wild species. There is a need to expand and refine existing methods, to improve specificity of action and effectiveness, and to reduce undesirable side-effects.

For human family planning, natural methods, despite their low efficiency, warrant a new look. The increased potential to offer accurate and practical kits for monitoring ovarian function may revive interest in the use of 'safe periods'. Further understanding of the mechanisms causing temporary infertility, prepubertally, during the period of suckling, and seasonally in some species, may lead to the development of more 'natural' intervention methods. To invoke ovulation suppression by the mechanisms present at suckling, without the mammary changes, may offer a control procedure without either short- or long-term side-effects.

Sheath barrier methods of contraception remain simple and essentially side-effect free, and have the added advantage of preventing the transfer of HIV and other venereal diseases. The addition of antiviral agents to both male and female sheaths makes them additionally useful in this regard. Sterilization is a very common fertility control measure worldwide. Demand for particularly female sterilization procedures may rise. An IUD offers perhaps a temporary alternative procedure. More simple and specific ways of interfering with sperm transport and maturation (within the female tract) and fertilization are possible. Probably only a very small number of sperm are normally permitted to reach the site of fertilization, and sperm–oocyte interaction includes mechanisms of both specific chemical attraction and sperm head membrane–zona pellucida binding prior to fertilization. The development of either specific antibodies or substances to limit single components of this system offer potential intervention mechanisms. Abortion may be performed less frequently or earlier, if contraceptive methods are more readily available and effective and if methods for the detection of congenital abnormalities could be applied sooner. Early abortion may be possible by the development of specific antagonists of the action of specific proteins of early pregnancy.

Since the initial development of orally active steroids, refinements have taken place in both their steroid content (with a reduction in or removal of the oestrogen component) and in the mode and means of administration. Injectables, slow release and biodegradable implants, and patches have been developed. The development and use of non-steroidal inhibiting hormones such as GnRH analogues, melatonin and its analogues, and inhibin analogues may prove advantageous. The use of such substances may depend on the development of methods to target repeated small quantities of active substance to a specific site.

The most successful steroid contraceptive methods have been developed for use in women. The development of methods for use in men has been limited by the close link between the control of spermatogenesis and libido in the male. The possible use of a testosterone derivative, in a protocol to block gonadotrophin support of both spermatogenesis and androgen production linked with the ability of the administered androgen to maintain libido but not maintain spermatogenesis, still awaits success. More specific methods may be to use inhibin-like substances to block only FSH secretion and hence spermatogenesis, or to specifically block spermiogenesis (the postmeiotic production of sperm from spermatids), or the maturation of sperm within the epididymis.

Family planning means more than contraception. Many more women, in particular, are seeking the opportunity to plan their families. Targets set by the UN Population Fund propose a 50% increase in family planning services.

Contraceptive methods developed for men or other male methods could be extensively applied as male sterilants for a variety of species and may also be one approach to developing a sterile-male protocol for mammalian pest control.

The manipulation of reproduction in man and in animals raises many ethical, social, and religious issues. The potential advantages and benefits of developments in this field are great, but there are very strongly held views for and against their use, and possibilities for abuse are also present. The rapid technological advances in the field of human reproduction have raised religious worries over contraception, ethical worries over abortion and embryo manipulation, and almost bewilderment over the possibilities for AI and IVF, donor gametes and surrogate uteri, providing 16 or more possible ways of 'making a baby'. The question has been raised as to whether people have the same rights to contraception and conception assistance as they have to food, shelter, and health. What is clear is that compared to efforts to reduce environmental degradation, reduction of population growth is achievable voluntarily. There is a demand for the prevention of unwanted births. The provision of contraceptives offers families this choice and hence the means to reduce population growth.

The raised awareness of people to environmental and welfare issues is increasingly being applied to domestic, wild, and endangered animals. Although there may be a feeling that certain manipulations such as cloning, though unthinkable for man, may be appropriately used in animals, there is a desire that where possible more natural, less invasive, non-drug based methods should be developed. There should be a move away from a relatively uncontrolled exploitation of wild and domestic species to a balanced, but not over-restricted, attitude in which progress, with safeguards, can be achieved.

Clearly challenges for reproductive biology remain. Further research and development will allow advances and benefits, without abuse, permitting more specific, safer, more effective and easy to use methods for contraception, family planning, and the alleviation of infertility in man; improved fertility, efficiency of production, and genetic potential in domestic animals; and possibilities for conservation and improved management of wild and endangered species.

Perhaps the ultimate contraceptive technology enables sex without reproduction and the ultimate conception assistance technology enables reproduction without sex, but the reproductive biologist has wider horizons.

Appendix: Glossary and abbreviations

Words in **bold** also have a separate entry; further information on topics can be obtained by reference to the index.

Abortion. Loss or expulsion of an **embryo** or **fetus** before independent existence is possible.

Abstinence. Periodic, regular or permanent avoidance of intercourse.

Accessory reproductive organs. Organs which nurture, transport and/or store **gametes** or **embryos** (e.g. **prostate, seminal vesicles, uterus, vagina**).

Accuracy. A validity criterion used to estimate the true level of a substance measured in an **assay**.

Acid phosphatase. See **phosphatase**.

Acromegaly. A **pituitary** disorder in man resulting in excess **growth hormone** secretion during adulthood.

Acrosin. An **acrosomal enzyme**; may be a good measure of the biochemical and physical integrity of the sperm acrosome.

Acrosome. The cap-like structure over the **head** of the **sperm**, containing **enzymes**.

Acrosome reaction. Morphological and chemical change in the **acrosome** membrane of the **sperm head** prior to **fertilization**.

ACTH. **Adrenocorticotrophin**; the **peptide hormone** of the **anterior pituitary** stimulating **corticosteroid** synthesis.

Activation. Of **sperm**, the increased activity that occurs just prior to **zona** penetration; or the change in the **zygote** signifying the onset of cell division.

Active immunity. That conferred by the administration of an **antigen** (or a relatively harmless form of it) to establish an adequate level of **antibody** and a primed population of cells to respond to the **antigen**.

Activins. **Hormones** of the **gonads** involved in the stimulation of **FSH** secretion and local **gonadal** control.

Adenoma. A benign epithelial tissue **neoplasm**.

Adenohypophysis. Consists of the **pars distalis, pars tuberalis** and pars intermedia: the first two are the **anterior pituitary** (see **pituitary gland** for full explanation).

Adenylate cyclase. Cell membrane **enzyme** complex catalysing the formation of **cyclic AMP**; many **protein/peptide hormones** operate *via* this system.

Adrenal cortex. Outer **steroid**-producing-portion of the mammalian adrenal gland.

Adrenarche. Maturation of the **adrenal cortex.**

Adrenocorticotrophin. ACTH.

Agonist. A substance that binds to a **receptor** and mimics the actions of the natural molecule (e.g. a **hormone**).

AI. Artificial insemination.

AID. Artificial insemination by **donor.**

AIDS. Acquired immune deficiency syndrome.

AIH. Artificial insemination by husband.

Aldehyde group. $-CHO$.

Aldosterone. The major **mineralocorticoid** of the **adrenal cortex** promoting sodium retention.

Alkaline phosphatase. See **phosphatase.**

Alpha configuration. See **configuration.**

Alpha-fetoprotein. AFP, measured in the **amniotic fluid** to aid in the diagnosis of **spina bifida.**

17α-Hydroxypregnenolone. Progestogen on the **steroid biosynthetic pathway** between **pregnenolone** and **dehydroepiandrosterone.**

17α-Hydroxyprogesterone. Progestogen on the **steroid biosynthetic pathway** between **progesterone** and **androstenedione**; parent compound of many synthetic **progestogens.**

α-N peptide. An N-terminal portion of **inhibin.**

α Subunit. A component of a **protein** containing two or more different **peptide** chains (e.g. **LH, FSH, inhibins, activins**).

Amenorrhoea. An absence of menstruation; may be **primary** if no cycles have occurred or **secondary** if cycles commence but subsequently cease.

Amine. Derivative of ammonia (NH_3) in which one or more hydrogens are replaced.

Amino acids. Compounds containing both an **amino group** and a **carboxy group**; the basic constituents of **proteins.**

Amniocentesis. Sampling of the amniotic fluid for the **diagnosis of congenital abnormalities** of the **fetus.**

Amino group. $-NH_2$.

Amniotic fluid. Fluid surrounding the **fetus** in the amniotic sac.

Ampullary glands. Paired male **accessory glands** in some species.

Analogue. A substance chemically related to the natural substance, may be an **agonist** or an **antagonist.**

Androgen binding protein. ABP, protein from the **Sertoli cells** that binds **androgens.**

Androgens. Sex **steroids** with 21 carbon atoms which have masculinizing properties; typically produced by the **testis.**

Andrology. The branch of medicine dealing with the functions of and diseases peculiar to the male sex organs.

Androstane. Hypothetical 'parent' compound of the **androgens** having 19 carbon atoms.

Androstenediol. The **androgen** derived from **dehydroepiandrosterone** in the **steroid biosynthetic pathway**.

Androstenedione. The **androgen** derived from 17α-hydroxyprogesterone in the **steroid biosynthetic pathway**.

Anoestrus. Period of **oestrous cycle** quiescence; may be prolonged in **seasonal breeders** and at **senescence**.

Anorexia nervosa. Syndrome involving a rigid restriction of food intake due to an abnormal fear of being fat.

Anovulation. Failure to **ovulate**.

Antagonist. Usually designates a substance that blocks the actions of **hormones** or **neurotransmitters** by **binding** to their **receptors**.

Anterior pituitary. Part of the **adenohypophysis** which includes the **pars distalis** and **pars tuberalis**; produces **hormones**, including **FSH**, **LH**, **ACTH** and **GH**.

Antiandrogen. An **androgen antagonist**.

Antibody. A **protein** produced by the body in response to a foreign substance, **an antigen**, to which it will specifically **bind**.

Antigen. A substance that activates the immune system, promoting **antibody** production.

Antihormones. **Antagonist** compounds that block the action of **hormones**, by **binding** to **receptors** and interfering with normal functions of the **hormone**.

Antiluteolysin. A **luteolysin antagonist**.

Antioestrogen. An **oestrogen antagonist**.

Antiprogestogen. A **progestogen antagonist**.

Antiserum. Serum containing **antibodies** with affinity for a specific **antigen**.

Antrum. The fluid-filled cavity of an **ovarian follicle**.

Aproned male. An intact male used as a **teaser** with a cover to prevent sperm transfer.

A ring. First ring of the **steroid** molecule.

Aromatase. **Enzyme** system catalysing the conversion of **androgens** to **oestrogens**.

Artificial insemination. Transfer of **sperm** to the female by means other than the male.

Artificial vagina. A device used for collecting **semen** from animals for **artifical insemination** and freezing.

Aspermia. Failure to produce **semen**.

Assay. See **bioassay**, **immunoassay**, **receptor** assay.

ATP. Adenosine triphosphate.

Atresia. A degradation which befalls all **oocytes** and **follicles** that do not **ovulate**.

Autocrine. Regulation of function by substances produced within the same cell type.

Azastene. A **3β-hydroxysteroid dehydrogenase** inhibitor, hence a **progesterone** synthesis inhibitor.

Azoospermia. Failure to produce **sperm**.

b. Prefix indicating a bovine (cattle) **hormone** (e.g. bTP).

Back pressure test. Applied to sows for **oestrus** detection.

Barrier methods. Contraceptive methods involving a physical block being established for **sperm** (e.g. **condoms, caps, diaphragms, sterilization**).

Base complementarity. Adenine:thymine; guanine:cytosine base pairing sequences.

BBT. Basal body temperature.

β Configuration. See **configuration**.

β Subunits. Components of a **protein** containing two or more different **peptide** chains (e.g. **FSH, LH, inhibins, activins**).

Bicycle system. The procedure of postponing **periods** for users of **combined oral contraceptive pills** by taking two packets in a row (i.e. two cycles before a 7 day break and a **period**).

Binding. Used in three contexts: the process of a **hormone** reacting with a **receptor**; the process of a **hormone** being joined to a serum **binding protein** (e.g. **SHBG**) for transport in the blood; the process of a **hormone** being joined to an **antibody** during an **immunoassay**.

Binding proteins. Proteins that attach to **hormones** for their transport in the blood.

Bioassay. The determination of a **hormone** by measuring its biological potency either *in vivo* or *in vitro*.

Biodegradable implants. Hormonal contraception implants that do not require removal when the **hormone** is exhausted.

Biogenic amines. Naturally occuring biologically active **amines**.

Biosensor. Device for the monitoring of physiolgical or biochemical functions of biological systems.

Biphasic pills. Combined oral contraceptives in which the ratio of **oestrogen:progestogen** varies to imitate the normal **menstrual cycle** profile in two phases.

Birth. The time or process of transfer of the **fetus** to be a free-living individual outside the mother's body.

Birth control. Limitation of the number of **births** by natural or artificial means; an alternative name for **family planning**.

Blastocyst. A stage of **embryonic cleavage** consisting of a fluid-filled ball of cells differentiated into an outer **trophoblast** layer and an **inner cell mass**.

Blastomeres. Cells produced by **cleavage** of the early embryo.

Blood–brain barrier. Resistance to transfer of molecules between the blood and the cerebrospinal fluid or **neurones**.

Blood–testis barrier. The barrier that blocks free exchange between the blood capillaries and fluid surrounding components of the **seminiferous tubules** other than **Sertoli cells** and **spermatogonia**.

Boar effect. The action of boar contact to induce **puberty** in prepubertal **gilts**.

Booroola gene. A gene of Booroola sheep imparting high **ovulation** rate.

Bound. See **binding**.

Breakthrough bleed. Unexpected bleed occurring between **withdrawal bleeds**, often whilst on a **contraceptive** regimen.

Breeding. The process of **mating**.

Breeding season. Season of the year when **mating** takes place.

B ring. Second ring of the **steroid** molecule.

Bromocriptine. A **dopamine agonist** used to inhibit **prolactin** secretion.

Bulbourethral glands. Cowper's gland, paired **accessory glands** in the male.

Buserelin. Highly potent **GnRH agonist**.

cAMP. Cyclic AMP, cyclic adenosine 3′,5′-monophosphate, a **second messenger** involved in **protein hormone** action.

Cancer. A malignant **neoplasm**.

Capacitation. The maturational change in the **sperm head** surface taking place in the female tract prior to **fertilization**.

Caps (cervical). Thimble-shaped **contraceptive** devices made to fit tightly around the **cervix** to prevent the passage of **sperm**.

Carcinogen. A substance that induces **cancer**.

Carbonyl group. $-C=O$.

Carboxyl group. $-COOH$.

Carnitine. Constituent of **semen** from the **epididymis**, in some species.

Cascade. Series of reactions, each one triggering the one that follows, often with an amplification of effect at each step.

Castration. Removal of **testes** or **ovaries** (the latter usually termed **ovariectomy**).

Catecholamines. Amines with 2-OH groups attached to the benzene ring (e.g. **dopamine, noradrenaline**).

Catecholoestrogens. Oestrogens with two $-OH$ groups attached to the **A-ring**.

Cervical mucus. Mucus produced by the **cervix**.

Cervix. The link between the **vagina** and the **uterus**.

Chernobyl. Town in the Ukraine, the site of the world's worst nuclear accident, April 26, 1986.

Chinball marker. Used to indicate that mating has occured in cattle.

Cholesterol. Sterol, precursor of **steroid** biosynthesis.

Chorion. An extra-embryonic membrane forming the outer cover around the **embryo**.

Chorionic gonadotrophins. Gonadotrophins derived from the chorion (e.g. **eCG, hCG**).

Chorionic villus sampling. Sampling of cells from chorionic villus of the **placenta** for genetical characterization for **congenital abnormality** assessment.

Chromosomes. Bodies in the cell nucleus containing the **genes**.

CIDR. Controlled internal drug release device; **vaginal** device used for **progesterone** slow release.

Citric acid. An organic acid, present in high concentration in the **semen** of most mammalian species.

Clearance. Usually refers to the removal of **hormone** or other substance from the blood (as in **metabolic clearance rate**).

Cleavage. Cell division of the early **embryo** following **fertilization**.

Clitoris. Erectile sex organ in the female, homologue of the **penis**.

Clomiphene. Nonsteroidal agent, with both **oestrogen** and **antioestrogen** activity, used in the treatment of human **polycystic ovarian disease**.

Clomiphene stimulation test. The use of **clomiphene** to test the integrity of the **hypothalamo–pituitary** axis. At least a doubling of **LH** and **FSH** 8 days after oral **clomiphene** suggests a normal axis.

Clone. Identical cells or individuals arising from a single cell or ancestor; nucleic acid sequences copied within a host cell during manipulation techniques.

Cloning. The process of reproducing identical copies of sequences of **DNA**, cells or individuals by other than natural means.

Cloprostenol. Long-acting $PGF_{2\alpha}$ **analogue**.

Coculture. Refers to the culture of different cell types together *in vitro* (e.g. coculture of **oocytes** with **granulosa cells** in **IVF** studies).

Cohort of follicles. Often used to refer to a group of growing **follicles** from which **preovulatory recruitment** occurs.

Coitus. Sexual intercourse by the insertion of the **penis** into the **vagina** as in mammals.

Collagenase. Enzyme involved in the breakdown of collagen, a process involved in **ovulation**.

Combined oral contraceptive pills. The 'standard' **contraceptive pill** containing **oestrogen** and **progestogen**, normally taken for 21 (or 22) days out of every 28.

Conception. The process of **fertilization** and the subsequent establishment of **pregnancy**.

Conceptus. The product of **conception**.

Condom. Barrier method of contraception using a sheath over the **penis**.

Configuration. Each substituent (e.g. OH group) on a steroid can project either above the plane of a ring, β-**configuration** (denoted by a solid line); or below the plane of a ring, α-**configuration** (denoted by a dotted line).

Congenital abnormality. An abnormality present at **birth**.

Conjugated equine oestrogens. The pregnant mare produces high levels of **oestrone sulphate**, and two **oestrogens** unique to equids, equilin and equilenin.

Conjugated steroids. Steroids converted into water-soluble products for excretion as **sulphates** or **glucuronides**.

Conservation. The act of keeping from damage or loss the flora, fauna and environment of the planet; the wise use of resources.

Contraception. The prevention of **fertilization** (and **pregnancy**) by reversible methods, but may also include **sterilization** and **abortion**.

Contraceptive pills. Oral contraceptives.

Copulation. Sexual union during which transfer of male **gametes** to the female occurs.

Corona radiata. The inner layer of **granulosa cells** surrounding the **ovum**.

Corpus luteum. Structure formed from the **follicle** after **ovulation**; secretes **progesterone** (and other hormones).

Cortical granules. Organelles of the **oocyte** contributing towards the **zona reaction**.

Cortical reaction. The loss of granules from the **oocyte** leading towards the **zona reaction**.

Corticosteroids. Steroids of the **adrenal** cortex.

Cortisol. The major **glucocorticoid** in many species.

Cowper's gland. See **bulbo-urethral gland**.

Creatine phosphokinase. An **enzyme** used as a **sperm mid-piece fertility** marker.

Creutzfeldt-Jakob syndrome. A disease of the brain, causing sponge-like degeneration.

CRH. Corticotrophin-releasing hormone, a **hypothalamic peptide** controlling **ACTH** release.

C ring. Third ring of a **steroid** molecule.

Cryoprotective agents. Substances (e.g. **glycerol, dimethyl sulphoxide**) which minimize the deleterious effects of freezing.

Cryptorchidism. Failure of the **testes** to descend into the **scrotum**.

Cumulus cells. See **cumulus oophorus**.

Cumulus oophorus. The **granulosa** cells surrounding the **oocyte**.

Curette. An instrument used to scrape the lining of the **uterus**.

Cyclic AMP. See **cAMP**.

Cushing's disease/syndrome. Disorders associated with excessive secretion of **glucocorticoids**.

Cyproterone acetate. An **antiandrogen**.

Cystic corpus luteum. A hollow fluid-filled **corpus luteum**.

Cystic follicle. Abnormal conversion of a **follicle** into a fluid-filled cyst.

D and C. Dilatation and curettage, an operation in which the **uterus** is dilated sufficiently to pass a **curette**, to scrape the lining.

Decidual reaction. Part of the **uterine** lining that becomes modified in response to **implantation**.

Deforestation. The removal and non-replacement of forest trees.

Dehydroepiandrosterone. DHA, **androgen** in the **steroid biosynthetic pathway**.

Delayed development. Phenomenon taking place in some bats in which **embryonic/fetal** development may proceed slowly.

Delayed fertilization. Phenomenon taking place in some bats in which **fertilization** by **sperm** stored in the **uterus** takes place in spring after hibernation.

Delayed implantation. Phenomenon taking place in several species in unrelated orders in which **blastocyst implantation** is delayed.

Depoprovera. **Medroxyprogesterone acetate**, a long-acting **progestogen contraceptive**, derivative of **17α-hydroxyprogesterone**.

Dexamethasone. Highly potent **glucocorticoid analogue**.

Dexamethasone suppression tests. Tests to determine the presence and origin of **Cushing's syndrome** (associated with elevated glucocorticoids). Elevated **cortisol** after overnight or low-dose **dexamethasone** suppression suggests **Cushing's syndrome**; elevated **cortisol** after high-dose **dexamethasone** suggests an adrenal **neoplasm** or **ectopic ACTH** production.

Diagnosis. The identification of a disease or syndrome by means of its symptoms.

Diaphragms. Circular shallow rubber domes with flexible rims which act as **contraceptives** by blocking the upper **vagina** and **cervix**.

Dibutyryl cyclic adenosine 3'5'-monophosphate. dbcAMP, an **analogue** of **cAMP** capable of rapidly entering cells and resisting degradation within the cell.

Dictyate stage. Prolonged 'resting phase' in oocyte **meiosis** which is terminated shortly before **ovulation**.

Diethylstilboestrol. DES, a potent non-**steroid oestrogen agonist**.

Dihydrotestosterone. Active form of **testosterone** in some target tissues.

Dimethyl sulphoxide. A **cryoprotective** agent.

Dinoprost. A **prostaglandin** $F_{2\alpha}$ **analogue**.

Diploid. State in which **chromosomes** occur as a homologous pair.

DNA. Deoxyribonucleic acid, contains the **genetic** information coded in specific sequences of its constituent nucleotides.

DNA probe. A defined nucleotide sequence used to detect a specific **DNA** sequence by base complementarity.

Domestic animals. Animals whose **breeding** is or can be controlled by humans.

Dominance. As applied in **ovarian follicles**, where one or more **follicles** are selected from a **cohort** of **follicles** to become the only developing **follicle**(s).

Donor. Refers either to the individual supplying **oocytes** or **embryos** for **embryo transfer** or **sperm** for **AI**, or to the **morula** from which cells are derived for **nuclear transfer.**

DOPA. Dihydroxyphenylalanine, a **dopamine** precursor which can cross the **blood–brain barrier.**

Dopamine. Catecholamine neurotransmitter, probably also a major **prolactin-inhibitory hormone.**

Doubling time. The time taken for a **population** to increase in size twofold.

Down regulation. A phenomenon of **receptor** control in which the sensitivity to **hormone** action is reduced by a reduction in **receptor** numbers, often induced by chronically elevated **hormone** concentrations.

Down's syndrome. A **congential abnormality** leading to physical and mental retardation resulting from a **trisomy** of **chromosome** 21.

D ring. Fourth ring of a **steroid** molecule.

Dwarfism. A stunting of growth, may be linked with a deficiency of **growth hormone** or **IGF-I.**

Dynamic testing. The active manipulation of the **hormonal** milieu by exogenous factors (drugs or **hormones** etc.) to evaluate more completely the potential responses of the **endocrine** system.

Dystocia. Painful or difficult **birth.**

e. Prefix referring to an equine (horse) **hormone** (e.g. **eCG**).

Early pregnancy factor. EPF, a factor produced very early after **fertilization.**

eCG. See **equine chorionic gonadotrophin.**

Ecology. The study of organisms in relation to their environment.

Ecosystem. A community of organisms interacting with each other and their environment.

Ectopic hormone production. Formation and release of **hormones** by tissues not usually associated with their production (e.g. **tumours**).

Ectopic pregnancy. A pregnancy site other than in the **uterine** cavity, usually the **fallopian tube.**

Efferent ducts of the testis. Structures linking the **seminiferous tubules** of the **testis** with the **epididymis.**

Ejaculation. The sudden expulsion of **semen.**

ELISA. Enzyme-linked immunosorbent assay, immunoassay system utilizing enzyme labels.

Embryo. The product of **conception**; term used for stages of intrauterine development until tissue and organ development is essentially complete, when it becomes a **fetus.**

Embryo splitting. The manipulative division of pre**implantation** stages of the **embryo.**

Embryo stem cells. Stem cells from the **inner cell mass** of the blastocyst.

Embryo transfer. The translocation of an **embryo** from another or the same individual or after **IVF** into the female reproductive tract (**oviduct** or **uterus**) for further development.

Endangered species. A species that is threatened with extinction.

Endocrine glands. Glands that secrete their products into the bloodstream.

Endocrine hypothalamus. All aspects of **hypothalamic** function involved in **hormonal** production or control.

Endocrinology. Study of **hormones**, their stimulation, production, actions and interactions.

Endogenous rhythm. Rhythm originating within an organism which persists though the external environment remains constant.

Endometriosis. A syndrome in which ectopic **endometrial** tissue 'bleeds' at each **menstrual period**.

Endometrium. The lining of the uterus, prepared for **embryo** attachment and **implantation**.

Enzyme. A **protein** catalyst promoting the rate of conversion of one substance into another.

Eosin. An orange dye, used in live:dead staining of **sperm**: viable **sperm** remain unstained, dead **sperm** take up the dye.

Epidermal growth factor. EGF, a poly**peptide** present and active in multiple tissue types including the **gonads**, closely related to **TGFα**.

Epididymis. Coiled tube connecting the **efferent ducts** of the **testis** to the **vas deferens**, within which **sperm** maturation and storage takes place.

Episodic. Pulsatile. Describes the intermittent nature of much hormone release, particularly the gonadotrophins as controlled by **GnRH** and the **pulse generator**.

Epostane. A 3β-hydroxysteroid dehydrogenase inhibitor, hence a **progesterone** synthesis inhibitor.

EPSI. Endometrial prostaglandin synthesis inhibitor.

Equine chorionic gonadotrophin. eCG, pregnant mare's serum gonadotrophin. Glycoprotein **hormone** produced by the **pregnant** mare, used as a natural **analogue** for **pituitary FSH** (and **LH**) for a long-acting **superovulatory hormone** in many species.

Erosion. Removal of top soil and vegetation by water, wind, ice and foot (animals and man).

Esters. Derivatives of acids.

ET. Embryo transfer.

Eutherian. Pertaining to mammals in which a well-formed **placenta** is present.

Extender. Used for diluting **semen** for short-term maintenance or freezing.

External genitalia. Externally visible reproductive organs.

Extra-gonadal. Factors (**gonadotrophins** etc.) other than those within the **gonads** that control those organs.

Extra-hypothalamic. Factors (**neurotransmitters etc.**) not produced by the **hypothalamus** that control the **endocrine hypothalamus**.

f. Femto, prefix before g or mol; 10^{-15}.

Fallopian tubes. Mammalian **oviducts**.

Family planning. Alternative term for **birth control**, implying that it is for a family to decide whether to limit its size.

Fecundity. The number of offspring actually produced.

Feedback control. Regulation exerted over **hormone** secretion by the product of a target gland; regulation exerted over an **enzyme-**controlled pathway by intermediates of the pathway. May be **positive** or **negative**.

Feedback loops. Pathways mediating **feedback control**. For **hormones, long loops** use the peripheral blood; **short loops** use the **hypo-physial portal** system; ultrashort (**paracrine**) loops and ultra-ultra **short-loops** (**autocrine**) act on neighbouring or the same cells respectively.

Female condom. Sheath for **barrier contraception** in the female.

Feral. **Domestic** or captive animal that has adopted a wild existence.

Ferning. Formation of a fern-like pattern when **cervical mucus** dries on a microscopic slide when collected during **oestrogen** dominance.

Fertility. The ability to produce offspring.

Fertilization. The union of the male and female **gametes**.

Fetoplacental unit. The joint involvement of the **fetus** and **placenta** for **steroid** synthesis and metabolism during **pregnancy** (the use of the term should not detract from the important maternal contributions to **steroid** metabolism).

Fetus. Conceptus (embryo) that has acquired essentially the morpho-logical characteristics of its species.

Fluorochromes. General name for fluorescent dyes; used for labelling **sperm** for separation and analysis.

Flushing. The process of improving nutrition to induce increased **ovu-lation** rate; the procedure of washing **embryos** from the female reproductive tract for collection; the vasomotor problem of the **menopause**.

Follicles. Structures in the **ovary** containing the developing **oocytes**.

Follicle pool. Usually refers to the pool of **primordial follicles** present in the **ovary** from **fetal** life; may sometimes refer to the **cohort of follicles** within the **ovary** from which preovulatory follicles are recruited.

Follicle-stimulating hormone. A glycoprotein hormone of the **anterior pituitary gland** involved in **gonadal** stimulation.

Follicular phase. The phase of an **oestrous/menstrual cycle** during which pre**ovulatory follicular** development occurs.

Follicular waves. The pattern of development of antral **follicles** which occurs in some species.

Folliculogenesis. The development of an **ovarian follicle** from a pri**mordial follicle** to the time of **ovulation**.

Follistatins. Members of a family of poly**peptides** (see also **inhibins, activins**) that modulate **FSH** release; may be more concerned as **binding proteins** of the **activins**.

Forward movement of sperm. Assessed *in vitro* as a parameter of motility which may be more compatible with **fertilizing** capacity.

Free. Used to refer to the non-**protein bound** fraction of a **hormone** (e.g. **steroid**) in the blood; also the fraction of **hormone** which is not **bound** in an **immunoassay** or **receptor assay**.

Freemartin. An intersexual **heifer** produced as a result of being a twin in **uterus** with a male calf.

Fructose. The main sugar contained in **semen** of many species.

FSH. Follicle-stimulating hormone.

GABA. γ-Amino-butyric acid.

Gametes. Ova and **sperm**.

Gametogenesis. Gamete formation: **spermatogenesis** and **oogenesis**.

γ-Amino-butyric acid. An **amine neurotransmitter**.

Genes. Hereditary factors carried on **chromosomes**.

Gene pool. All the **genes** present in an interbreeding **population** at a particular time.

Genetic diversity. Natural variation in **genes** within a **population** enabling change to occur by natural selection or other means.

Genetic engineering. A change in the genetic constitution of an organism brought about by a directed manipulative technique.

Genetics. The study of **genes** and their effects.

Gene transfer. The addition of **DNA**, allowing the replacement of a missing gene, or the introduction of an exotic gene in a manner that allows it to be reproduced.

Genome. The total **genetic** material within a cell or individual.

Genotype. The **genetic** make-up of a cell or organism.

Gestation period. Pregnancy: time required to complete normal prenatal development, between **conception** and **birth**.

GH. Growth hormone.

GIFT. Gamete intrafallopian **transfer**.

Global warming. The heating up of the earth due to the greenhouse effect.

Glucocorticoids. Corticosteroids involved in carbohydrate metabolism.

Glucuronides. (Steroid), see **steroid glucuronides**.

Glycerol. Used as a **cryoprotectant**, to prevent ice damage of cells during freezing.

Glycoprotein hormones. Protein hormones containing carbohydrate (sugar) groups (e.g. **FSH, LH, eCG, hCG**).

Glycosylation. The addition of sugar residues to a molecule (e.g. to a **protein** to form a **glycoprotein**).

GnRH. Gonadotrophin-releasing hormone.

GnRH-associated peptide. GAP, a **peptide** derived from **prepro-GnRH**.

GnRH stimulation test. Tests the ability of the **anterior pituitary** to secrete **gonadotrophin**, by measuring the **gonadotrophin** response to exogenous **GnRH** challenge.

Gonadostat. Usually applied to the **hypothalamo–pituitary negative feedback** response system, the high sensitivity of which to **gonadal steroids** maintains minimal **gonadal** activity during the prepubertal period.

Gonadotrophin-releasing hormone. Decapeptide of the **hypothalamus** which stimulates **gonadotrophin** synthesis and secretion.

Gonadotrophins. Include the **anterior pituitary gonadotrophins** (**FSH** and **LH** that stimulate the **gonads**) and the **chorionic gonadotrophins**.

Gonads. Organs specialised to produce **gametes** (the **testes** and **ovaries**), also produce **steroid** and **peptide hormones**.

G protein. Protein(s) coupling the **receptor** to **adenylate cyclase**.

Gossypol. Component of some cotton seed oil, possible male **contraceptive**.

Graafian follicle. Mature **ovarian follicle** prior to **ovulation**.

Granulosa cells. Ovarian **follicular** cells surrounding an **oocyte**.

Greenhouse effect. Global warming due to retention of atmospheric heat caused by the action of **greenhouse gases**.

Greenhouse gases. Especially carbon dioxide, also chlorofluorocarbons, methane and nitrous oxide.

Green issues. A variety of scientific, social and environmental issues (e.g. conservation, the use of renewable energy sources, sustainable agricultural systems etc.).

Growth factors. A range of regulators affecting growth, differentiation and proliferation (e.g. **TGF, IGF-I**).

Growth hormone. Protein **anterior pituitary hormone** affecting growth and metabolism.

Gynaecology. The branch of medicine dealing with functions and diseases peculiar to women.

h. Prefix referring to a hormone of human origin (e.g. **hCG**).

Half-life (biological). The time taken for half of a substance (e.g. **hormone**) secreted or administered to be lost from the body.

Haploid. In which the **chromosomes** are represented unpaired; the number characteristic of **gametes**; half the **diploid** number.

hCG. See **human chorionic gonadotrophin**.

Head. The front portion of a **sperm**.

Heat. See **oestrus**.

Hemizona assay. The use of half **zonae pellucidae** to assess **sperm** binding, as a method of predicting the **fertilizing** potential of **sperm**.

Hibernation. A dormant state, involving decreased metabolism, in which some animals overwinter.

Hirsutism. A condition characterized by growth of hair in unusual places or in unusual amounts.

HIV. Human immunodeficiency virus, causes **AIDS**.

hMG. **Human menopausal gonadotrophin.**

Homeostasis. The tendency for a balanced internal state within the body, **hormone** homeostasis often maintained by **negative feedback**.

Hormone. A chemical messenger of the **endocrine** system: in this book also more generally applied to **neuroendocrine, paracrine, autocrine** and **intracrine** factors.

Hormone assay. The measurement of the concentration of a **hormone** in a tissue, body fluid or extract.

Hormone receptor. Specific cellular **hormone binding** component.

Hormone replacement therapy. See **HRT**.

Hot flushes. A common vasomoter problem associated with the **menopause**.

HRT. Hormone replacement therapy: the replacement of **oestrogen** (and **progesterone**) in women to prevent **menopausal** symptoms.

Human chorionic gonadotrophin. hCG; the **glycoprotein hormone** produced by the **trophoblast** cells of the early human **embryo**; involved in maternal **pregnancy recognition**; maintains **corpus luteal** function in early **pregnancy**.

Human menopausal gonadotrophin. See **menopausal gonadotrophin**.

Human placental lactogen. See **placental lactogen**.

Hybrid vigour. **Genetic** superiority due to the presence of hetero-zygosity (possessing a dominant and a recessive **gene**) in a number of different **gene** pairs.

Hydroxylation. The introduction of an $-OH$ group onto a molecule (steroid).

Hydroxyl group. $-OH$.

Hydroxysteroid dehydrogenases. Enzymes which catalyse the oxi-dation or reduction of 'oxy' functions (e.g. 3β, 17β).

5-Hydroxytryptamine. Serotonin; an **indoleamine neurotransmitter**.

Hyper-. Prefix meaning excess.

Hyperactive sperm. The ability of **sperm** to express hyperactivation, a high-amplitude or looping motion, is closely correlated with the ability to **fertilize oocytes**.

Hyperprolactinaemia. High blood **prolactin** concentration.

Hyperspermia. High **semen** volume (greater than 5.5 ml in man).

Hypertension. High blood pressure.

Hypo-. Prefix meaning either deficient (e.g. **hypogonadism**) or under (e.g. **hypothalamus**).

Hypogonadism. A deficiency of **gonadal** function.

Hypogonadotrophic hypogonadism. A deficiency of **gonadal** function due to a deficiency of **gonadotrophic** function.

Hypo-osmotic test. A test involving water uptake by **sperm** which examines the integrity of the **sperm** membrane.

Hypospermia. Low **semen** volume (less than 1.5 ml in man).

Hypophysectomy. Removal of the **pituitary gland**.

Hypophysial–portal system. The portal (capillaries at each end) blood system linking the **median eminence** with the **anterior pituitary gland**.

Hypophysis. Pituitary gland.

Hypothalamo–pituitary–gonadal system. The integrated control system involving **hypothalamic GnRH, pituitary gonadotrophins, gonadal hormones** and the **feedback loops** involved in the control of **gonadal** function.

Hypothalamus. The portion of the brain involved in numerous physiological control mechanisms, including **hormonal** control of **anterior pituitary** function.

Hysteroscopy. Direct examination of the **uterine** cavity.

Hysterosalpingography. X-ray of the **uterus** and **fallopian tubes**, allowing views of **uterine** cavity and **tubal** patency.

ICI. Intra-**cervical insemination**.

IGF-I. See insulin-like growth factor I.

Immune response. Production of **antibodies** in response to **antigens**.

Immunity. Protection against a foreign substance afforded by previous exposure to that substance (**active immunity**) or the transfer of preformed **antibody** (**passive immunity**).

Immunization. The act or process of conferring **immunity**.

Immunoassay. An assay using an **antibody** as a specific **binding** reagent.

Immunocastration. The **immunization** against a component of the **testicular** control system, producing inactive **testes**.

Implant. Refers to any **hormone** capsule, with or without a degradable or nondegradable matrix or container (e.g. **contraceptive steroid, melatonin** etc.).

Implantation. The process of interaction and attachment of the **blastocyst** with the **endometrium**.

Impotence. Lack of sex drive and inability for sexual intercourse.

Inbreeding. Mating of **genetically** similar individuals.

Indoleamines. Amines containing a benzene nucleus combined to a pyrole ring.

Induced luteal phase. The induction of a **luteal phase** (**pseudopregnancy**) by a sterile **mating** (e.g. mouse).

Induced ovulation. The induction of **ovulation** by stimuli associated with **mating** (e.g. rabbit).

Infertility. The inability to produce viable young within a stipulated time characteristic of each species.

Inhibins. Peptide hormones of **Sertoli** and **granulosa** cells associated with the inhibition of **FSH** release and local **gonadal** control.

Inner cell mass. The early **embryonic** cells in the **blastocyst.**

Insemination. The introduction of **sperm** into the female reproductive tract (or elsewhere) by natural or artificial means for **oocyte fertilization.**

Insulin-like growth factor I. IGF-I, **peptide** with a close structural and functional similarity to insulin; produced by the liver as a mediator of **GH** action, also produced by other tissues including the **ovary** where it is involved in a number of **intra-ovarian** control mechanisms.

Interferons. Proteins induced in an animal in response to a viral challenge; include the **trophoblast proteins.**

International standards. Standard preparations (e.g. of **hormones**), used for standardization of assays.

Interrupted coitus. Variations on **coitus**, preventing **sperm** transfer to the woman: interuptus – withdrawal prior to emission; reservatus – resistance of orgasm; saxonis – diversion of **sperm** into the bladder.

Intracrine control. Control system in which the product of a cell controls the same cell without being released.

Intra-ovarian control. Control systems for the **ovary** originating within the **ovary**; may be **autocrine, paracrine** or **intracrine.**

Intra-testicular control. Control systems for the **testes** originating within the **testes**; may be **autocrine, paracrine** or **intracrine.**

Intrauterine device. IUD, a small device inserted into the **uterus** to prevent **pregnancy**; may be called the coil or the loop.

Inverdale gene. A gene of Inverdale sheep in which the heterozygote has a high **ovulation** rate.

In vivo. 'In life', refers to studies on the whole animal.

In vitro. 'In glass', refers to studies on living material performed outside the body.

In vitro **fertilization.** IVF, **oocyte fertilization** outside the body.

IPI. Intraperitoneal **insemination: insemination** into the peritoneal cavity.

IRP. International reference preparation: a preparation distributed internationally for standardizing assays without **international standard** status (e.g. IRP/hMG).

iu. International unit, accepted unitage of a substance (e.g. **hormone**) in terms of an **international standard** or reference preparation.

IUD. See **intrauterine device.**

IUI. Intrauterine **insemination: insemination** into the **uterus.**

IVF. *In vitro* **fertilization.**

IZF. Intrazonal **fertilization**: injection of a **sperm** directly into the **perivitalline space.**

Karyoplast. Nucleus surrounded by a small amount of cytoplasm and cell membrane.

Karyotype. The **chromosomal** constitution of a cell (or individual).

Klinefelter's syndrome. An **infertile** condition linked with an XXY **trisomy**.

Labia. Folds of skin on each side of the **vulva**.

Lactation. Production of milk from **mammary gland**.

Lactation amenorrhoea. Lack of human female (primate) **menstrual cyclicity** caused by **lactation/suckling**.

Lactation anoestrus/anovulation. Lack of **oestrus** and **ovulation** induced by **lactation/suckling**.

Laparoscopy. Examination of the abdominal cavity with a tiny viewing tube.

Laparotomy. Incision through the abdominal wall.

Leucocytes. White blood cells.

Levonorgestrel. **Progestogen** derived from **19-nortestosterone** used in **Norplant, contraceptive implants** and **progestogen**-containing **IUDs**.

Leydig cells. The interstitial cells of the **testes**, lying between the **seminferous tubules**; secrete **androgens**.

LH. Luteinizing hormone.

LH surge. The sudden secretion of **LH** taking place during the pre**ovulatory** period leading to **ovulation, oocyte maturation** and **luteinization**.

Libido. The urge associated with sex drive.

Life expectancy. The average time to death at a particular age under particular conditions (lifestyle, country etc.).

Logit. A method of transforming data derived from **radioimmuno-** and other **assays** to linearize a sigmoid curve (logit $Y = \log_e(Y/1 - Y)$).

Long-day breeder. A **seasonal breeding** species that breeds in the spring.

Long-loop feedback. Usually refers to **hormonal feedback** linking the target gland (**gonad**) with the **hypothalamo–pituitary** unit.

Luprostiol. A $PGF_{2\alpha}$ agonist.

Luteal phase. The phase of an **oestrous/menstrual cycle** during which the **corpus luteum** is active.

Luteinization. Morphological and **steroidogenic** (biochemical) changes associated with the conversion of the **follicular** remnants (after **ovulation**) into a **corpus luteum**.

Luteinizing hormone. A **glycoprotein hormone** of the **anterior pituitary** gland involved in **gonadal** stimulation.

Luteolysin. A substance that promotes the regression of the **corpus luteum**.

Luteolysis. The functional and morphological degeneration of the **corpus luteum**.

Luteotrophin. A substance that maintains the structure and function of the **corpus luteum**.

m. Milli, prefix before g, mol or unit, 10^{-3}.

μ. Micro, prefix before g or mol, 10^{-6}.

Maklar slide. Special microscopic slide for **sperm** counting and observing **sperm** motility.

Malignant (tumour). Spreads to other parts of the body.

Mammary glands. Milk producing glands of mammals.

Marsupials. Mammals characterized by the posession of a pouch in which young are carried after **birth** in an undeveloped condition.

Mating. Usually refers to the act of **copulation**.

Median eminence. Region of the **hypothalamus** where the **peptidergic neurone** terminals release **hormones** into the upper end of the **hypophysial–portal system**.

Medroxyprogesterone acetate. Depot synthetic **progestogen** derived from **17α-hydroxyprogesterone**.

Megazoo. An extended zoo or wildlife park.

Meiosis. The reduction division of the cell nucleus to produce **haploid** cells.

Melatonin. N-acetyl 5-methoxytryptamine; **pineal hormone** involved in **seasonal** rhythms.

Menarche. Initiation of **menstrual cycles**.

Menopausal gonadotrophin. Usually describes **gonadotrophin** extracted from the urine of post**menopausal** women.

Menopause. Cessation of menstrual cycles.

Menses. The discharge of blood from the degenerating **uterus** lining in a non-**pregnant** higher primate cycle.

Menstrual cycle. Uterine cycle of a non-**pregnant** higher primate involving a menstrual bleed (**menses**).

Mestranol. An **oestrogen analogue**.

Metabolic clearance rate. MCR, the volume of blood completely cleared of **hormone** per unit time.

Methalibure. Drug used for synchronizing pig **oestrous cycles**; withdrawn extensively as was found to be a **teratogen**.

Methyl testosterone. A substituted **testosterone** derivative.

Microadenoma. A small benign **tumour** of glandular origin, often the source of **prolactin** in cases of **hyperprolactinaemia**.

Midpiece. The middle portion of a **sperm**.

Mifepristone. RU 486.

Mineralocorticoids. Corticosteroids involved in mineral metabolism.

Minipill. Female **oral contraceptive** containing only low amounts of **progestogen**.

Minipump. A device for the continuous or **episodic** delivery of small quantities of **hormone** (or drug) *in vivo*.

Mitosis. The division of the cell nucleus resulting in the distribution of a complete set of **chromosomes** to each daughter cell.

Mittelschmerz. Mid-cycle abdominal pain occurring in some women, perhaps associated with **ovulation**.

MOET. **Multiple ovulation** and **embryo transfer**, the composite procedure involving **follicular** stimulation, **ovulation**, **AI**, and **embryo transfer**.

Monoclonal antibody. Antibody produced from a single **clone** of cells *in vitro*.

Monoculture. The cultivation of large stretches of a single crop which are harvested all at once.

Monoestrus. Displaying a single **oestrous cycle**: as in **seasonally monoestrous**, displaying one **oestrous cycle** with each **breeding season**.

Monogamy. A single mate.

Monosomy. A condition in which one member of a specific **chromosome** pair is missing.

Monotocous. Giving **birth** to a single offspring.

Monotremes. Egg-laying mammals.

Morning-after pill. **Post-coital contraceptive**.

Morning sickness. Nausea or vomiting occurring on rising, an early symptom of **pregnancy**.

Morula. An early **embryo** consisting of a solid ball of cells, becomes a **blastocyst**.

Mouse uterus test. A **bioassay** used for the measurement of total **gonadotrophin** activity: involves studying the increase in **uterine** weight of immature female mice.

Mucus. A slimy fluid produced, for example, by the **cervix**, i.e. **cervical mucus**.

n. Nano, prefix before g or mol, 10^{-9}.

Naloxone. An opiate **receptor antagonist**.

Naloxone stimulation test. The use of the **opiate** receptor **antagonist** to test for an **opiate** block of the **pulse generator**.

Naltrexone. An **opiate receptor antagonist**.

Naphthelene derivatives. Term used for derivatives of naphthalene (which contains two condensed benzene rings) which are analogues of **melatonin**.

Negative feedback control. A process whereby a **hormone** produced by a target organ inhibits the stimulatary mechanisms for that organ (e.g. **gonadal steroids** inhibiting **gonadotrophin** secretion).

Neonate. Newborn; also designates changes occuring after **birth**.

Neoplasm. Benign or **malignant tumour** associated with cell proliferation.

Neural feedback. **Feedback** to the **hypothalamus via** neural pathways (e.g. as in **induced ovulators**).

Neurocrine control. Control mechanisms where the product of a nerve cell (**neurotransmitter**) acts on another cell.

Neuroendocrine control. The **hypothalamic** control of the **anterior**

pituitary where the product of a nerve cell (**peptidergic neurone**) is transported **via** the **hypophysial–portal** system to stimulate or inhibit **anterior pituitary hormone** synthesis and secretion (e.g. **GnRH** stimulation of **gonadotrophin** synthesis and release).

Neurohypophysis. Posterior pituitary gland.

Neurone. Major cell type of the nervous system specialized for information transmission.

Neurotransmitters. Regulators produced by **neurones** to transmit impulses across synapses.

Non-return. The failure to observe **oestrus** at the expected time; used to diagnose **pregnancy** in **inseminated** animals.

Noradrenaline. Norepinephrine, a **catecholamine neurotransmitter** and product of the adrenal medulla.

Norethindrone. See **Norethisterone**.

Norethisterone. A **progestogen** derived from **19-nortestosterone**, used as an **oral contraceptive**.

Norethisterone acetate. A **progestogen** derived from **19-nortestosterone**, used as an **oral contraceptive**.

Noristerat. A reservoir **progestogen contraceptive**.

Norplant. A **silastic** reservoir filled with **progestogen, levonorgestrel**, used as a long-acting **contraceptive**.

19-Nortestosterone. **Testosterone** without the methyl group at position 10; parent compound for many synthetic **progestogens**.

Nuclear transfer. Transplantation of a cell nucleus to another cell.

Nymphomania. Uncontrolled sexual desire or **oestrus** behaviour.

o. Prefix refers to a **hormone** of ovine (sheep) origin (e.g. **oLH**).

Oestradiol-17β. A naturally occurring **oestrogen** containing two -OH groups, the main secretory product of the **ovarian follicles** in most species.

Oestradiol benzoate. The benzoic acid ester of **oestradiol**.

Oestradiol stimulation test. The use of **oestradiol** administration to stimulate **LH** release, to test for an intact **oestrogen positive feedback** response.

Oestrane. Hypothetical 'parent' compound of the **oestrogens** having 18 carbon atoms.

Oestriol. A naturally occuring **oestrogen** containing three -OH groups; main **oestrogen** present during human **pregnancy**.

Oestrogens. Female sex **hormones** with 18 carbon atoms; named as the **hormones** bringing animals onto **oestrus** or **heat**.

Oestrone. A naturally occuring **oestrogen** containing one -OH group.

Oestrous cycle. The regular occurence of **oestrus** and related **ovarian**, behavioural and reproductive tract phenomena associated with non-**pregnancy** in many mammals.

Oestrus. Heat, the state in which a female mammal will stand voluntarily to be mounted by the male.

Oligozoospermia. A low **sperm** count (in man of less than 20×10^6/ml).

Oocytes. Female germ cells, may be **primary** or **secondary**.

Oocyte maturation. Meiosis of **oocytes**; commences in **fetal** life and is only completed at **fertilization**.

Oocyte recovery. Usually refers to the collection of **oocytes** from **ovarian follicles** either *in vitro* or *in vivo* for use in **IVF** systems.

Oogenesis. The process of production of female **gametes**.

Oogonia. Diploid ovarian cells, present only during fetal life, divide **mitotically** to form more **oogonia** and subsequently mature to form **primary oocytes**.

Ooplasm. The cytoplasm of the egg.

Opiate peptides. Include endorphins and encephalins; act on receptors affected by morphine.

Opioids. See **opiate peptides**.

Oral contraceptives. Synthetic orally active **oestrogens** and **progestogens** used in fertility control.

Ovariectomy. The removal of the **ovaries**.

Ovary. The female **gonad**.

Oviducts. Fallopian tubes, the site of **fertilization** and early **cleavage** stages of the **embryo**.

Ovulation. Release of the **oocyte** from the **ovary**; may be **induced** or **spontaneous**.

Ovum. Female gamete, or egg.

Oxytocin. Neurohypophysial **peptide** involved in milk let down and **parturition**.

p. Prefix refers either to a **hormone** of porcine (pig) origin (e.g. **pFSH**), or to pico before g or mol, 10^{-12}.

Paracrine control. Control in which a **hormone** released by one cell type acts locally to control another cell type.

Parallelism. A validity criterion for **assays**, in which the similarity of slope of the **assay** curves for the standard and unknown samples is checked.

Pars distalis. Component of the **adenohypophysis**.

Pars tuberalis. Component of the **adenohypophysis**.

Parturition. Birth; can also refer to the process of labour.

Passive immunity. That conferred by the administration of preformed **antibody**.

Patches. A method of administering a **hormone** (e.g. **oestrogen**) by the use of impregnated patches stuck on the skin.

Penis. The male sexual organ of copulation.

Peptide. A compound consisting of a chain of **amino acids** (e.g. deca-**peptide** – ten **amino acids**; poly**peptide** – many **amino acids**).

Peptidergic neurones. Hypothalamic cells that synthesize and secrete peptides (e.g. **GnRH** cells).

Peri-. Prefix meaning 'around' (e.g. perinatal, around **birth**; perimenopausal, around the **menopause**).

Period. Usually refers to the time of the **menses** during a **menstrual cycle**.

Peritubular cells. The cells surrounding the **seminiferous tubules** of the **testes**.

Perivitalline space. The space between the **zona pellucida** and the **vitalline membrane** of the **oocyte**.

Persistent corpus luteum. A **corpus luteum** remaining active beyond its normal life span, usually as a result of a **uterine** lesion.

Pest. A species whose existence conflicts with human profit, welfare or convenience.

PG. Prostaglandin (e.g. $PGF_{2\alpha}$).

pH. A scale of acidity (1–7) and alkalinity (7–14).

Phenotype. The characteristic form of an individual, determined by the interaction during development between the **genotype** and the environment.

Pheromones. Substances secreted by an organism into the environment which act on other members of the same species.

Phosphatase. An **enzyme** able to split phosphate groups from organic molecules which operates in either acid (**acid phosphatase**) or alkaline (**alkaline phosphatase**) medium; important constituents of **semen** in some species.

Photoperiod. Duration of the light phase of a recurring light–dark cycle.

PID. Pelvic inflammatory disease.

'Pill'. Usually designates an **oestrogen** plus **progestogen** oral **contraceptive**.

Pineal gland. The **endocrine gland** in the head which secretes **melatonin**.

Pituitary gland. Or **hypophysis**; consists of **anterior** and **posterior** lobes and the **adenohypophysis** and **neurohypophysis** in a rather confused terminology thus:

Adenohypophysis $\begin{cases}\end{cases}$	**Pars distalis** $\Big\}$ **Pars tuberalis**	**Anterior pituitary**
	Pars intermedia $\Big\}$	**Posterior pituitary**
Neurohypophysis $\begin{cases}\end{cases}$	Neural lobe $\Big\}$ Infundibular stem $\Big\}$ **Median eminence**	Infundibulum

The terms **anterior** and **posterior pituitary** are usually used, except when specific reference is necessary.

Placenta. Feto–maternal organ for the transfer of gases, nutrition and waste in **eutherian** mammals.

Placental lactogen. **Hormone** produced by the **placenta**, in the human involved in fat utilization.

Placental retention. The failure of a **placenta** to be shed.

Plasmin. Implicated in **ovulation**.

Plasminogen. Plasmin precursor.

Plasminogen activator. Involved in the conversion of **plasminogen** to **plasmin**.

Platelet activating factor. PAF, produced by pre**implantation embryos** of some species.

P Mod S. Protein modulating **Sertoli cell** function, a **paracrine** factor of the **peritubular cells** of the **testis** which regulates the **Sertoli cells.**

PMSG. Pregnant mares' serum gonadotrophin.

Polar body. Small cell consisting almost entirely of nuclear material formed during the division of **primary** and **secondary oocytes.**

Polyclonal antibody. Antibody arising from more than one cell or **clone.**

Polycystic ovaries. Ovaries containing large fluid-filled **cystic follicles** as in the **Stein–Leventhal** syndrome.

Polygamy. More than one mate (polyandry, one female with many males; polygyny, one male with many females).

Polymerase chain reaction. Multiplication of a specific **DNA** fragment using repeated cycling of the **enzyme** system for **DNA** synthesis.

Polyoestrus. Having more than one **oestrous cycle** per **season** or year.

Polysaccharides. A group of long-chain carbohydrates.

Polyzoospermia. A **sperm** count in man of greater than 250×10^6/ml.

Polyspermy. More than one **sperm** entering the **perivitalline space** for **fertilization.**

Polytocous. Giving **birth** to multiple young.

Population. All members of the same species living in a defined area.

Population control. Often used as an alternative to **family planning.**

Population explosion. Often refers to the huge increase in human **population** over recent decades.

Positive feedback control. Feedback in which there is a stimulation of increased secretion of the stimulatory gland by its target gland product (e.g. **oestrogen** from **preovulatory follicles** stimulating the **LH surge**).

POST. Peritoneal **oocyte** and **sperm transfer.**

Post-coital contraceptives. Contraceptives that act after **sperm** have entered the female reproductive tract.

Post-coital test. The examination of **cervical mucus** after **coitus** to define abnormalities of **sperm** function.

Posterior pituitary gland. Includes the **neural lobe** and the pars intermedia of the **pituitary gland.**

Precision. A validity criterion used in **assays** to estimate the errors associated with the measurement.

Pregnancy. Gestation, the period from **conception** to **birth.**

Pregnancy recognition. The process of signalling between the developing **embryo** and the mother for **corpus luteal** function to be extended and the **pregnancy** to be maintained.

Pregnancy tests. Methods used to diagnose **pregnancy**.

Pregnane. Hypothetical 'parent' compound of the **progestogens** having 21 carbon atoms.

Pregnant mares' serum gonadotrophin. PMSG, see **equine chorionic gonadotrophin**.

Pregnenolone. **Cholesterol**-derived precursor of **progesterone** and other **steroids**.

Preovulatory surge. The sharp output of **LH** (and **FSH**) which induces **ovulation**.

Prepro-GnRH. Large **peptide** which contains **GnRH**.

PRID. **Progesterone**-releasing intra**vaginal** device.

Primary follicle. Consists of an **oocyte** surrounded by a few layers of **granulosa cells**.

Primary oocytes. **Diploid** cells, differentiated from **oogonia**, form **secondary oocytes** during **oogenesis**.

Primary spermatocytes. **Diploid** cells, differentiated from **spermatogonia**, form **secondary spermatocytes** during **spermatogenesis**.

Primordial follicle. Consists of an immature **oocyte** surrounded by a single layer of flattened pre**granulosa cells**; form the **follicular pool** from which **folliculogenesis** takes place.

Progestogens. Progestins or progestagens; natural or synthetic **steroids** with **progesterone**-like actions which maintain **pregnancy**.

Progestogen alone pills. Mini-pills; **oral contraceptives** containing only a **progestogen** in low doses.

Progestogen withdrawal test. The induction of a **vaginal bleed** following **progestogen** injection.

Progesterone. Principle naturally occurring **progestogen** in all species.

Prolactin. The **protein hormone** of the **anterior pituitary** involved in the stimulation of milk secretion; has certain **gonadal** control properties in some circumstances.

Prolactin inhibitory and releasing factors. Various **hypothalamic** factors have been implicated in the control of **prolactin** release including **dopamine** as an inhibitory factor.

Pronuclei. (Male and female), the two sets of **haploid chromosomes** surrounded by membranes which become visible after **oocyte – sperm** fusion.

Prostaglandin $F_{2\alpha}$. $PGF_{2\alpha}$, the **luteolytic** agent in several species; involved in **parturition**; also probably plays a role in **ovulation**.

Prostaglandins. Group of lipid-soluble acids with various endocrine functions. The capital letter (A, E, F) defines the ring structure, the subscript (1, 2) indicates the number of double bonds (e.g. PGE_2).

Prostaglandin synthetase. Enzyme involved in synthesis of **prostaglandins**.

Prostate. A male **accessory gland**.

Proteases. Enzymes that cleave **proteins**.

Proteins. Complex molecules composed of linked **amino acid** units.

Protein-bound. The attachment of **hormones** to **proteins** for transport in the blood (e.g. **SHBG**).

Pseudopregnancy. Pregnancy symptoms without **pregnancy**, usually linked with the secretion of progesterone; induced in some species (e.g. mouse) by **mating** with a **sterile** male.

Puberty. Changes associated with sexual maturation (e.g. ability to produce **sperm** or **ova**).

Pulse generator. The components of the **endocrine hypothalamus** and its inputs leading to the **pulsatile** output of **GnRH** and **gonadotrophins**.

Pulsatile. Episodic; the intermittent nature of the secretion of many **hormones**.

Quinacrine. Originally a malaria suppression drug; a possible non-surgical female sterilant.

Radioactive iodine. Radioisotope of iodine used to label **proteins** for **radioimmunoassays**, usually ^{125}I.

Radioimmunoassay. Assay method based upon the competition between a **radioisotope**-labelled substance (e.g. **hormone**) and an unlabelled **hormone** for a specific **antibody**.

Radioisotope. Radioactive form of an element.

Ram effect. The induction of **ovulation** following association of an **anoestrous** anovular ewe with a ram, probably mediated by a **pheromone**.

Receptor. Cell component **binding** a **hormone** (or other regulator) leading to the changes in cell function which are the 'responses' to the **hormone**.

Receptor assay. Similar to **RIA**, except that specific biological **receptors** or **binding proteins** are used rather than **antibodies**.

Recipient animal. A female that receives an **embryo** at **embryo transfer**.

Recipient oocyte. An **oocyte** receiving a transferred nucleus or other genetic material.

Recombinant DNA. The combination of two **DNA** sequences of different origin by means of **genetic engineering**.

Recombinant DNA technology. Genetic engineering.

Recruitment of follicles. The processes whereby either **follicles** leave the **follicular pool**, or are stimulated for continued development up to **ovulation** by **gonadotrophins**.

Reflex ovulator. See **induced ovulator**.

Relaxin. A **peptide hormone** with different sites of synthesis (including **corpora lutea** and **testes**) and actions (including pre**parturition cervical** softening and dilation).

Releasing hormones. **Hormones** synthesized by **hypothalamic neurones**, stimulate **anterior pituitary hormone** secretion (include **GnRH** and **CRH**).

Reproduction. The process of replication of individuals.

'Rhythm' method of contraception. A **contraceptive** method based upon being able to monitor the **safe period** of the **menstrual cycle** by checking **cervical mucus**, **basal body temperature** and cycle length.

RIA. Radioimmunoassay.

RU 486. Mifepristone. An **antiprogestogen** used for **pregnancy** termination.

Safe period. The period of time during a **menstrual cycle** during which a woman can have sexual intercourse and not expect to conceive.

Scrotum. The sac into which the **testes** descend in many mammals.

Season. The temporal (e.g. spring, autumn) or environmental (dry, wet) period of the year; also used to refer to the effects of season on **reproduction** (e.g. **breeding season**).

Second messenger hypothesis. The mechanism of **hormone** action whereby the first messenger (**hormone**) action *via* cell plasma membrane **receptors** is mediated by an intracellular second messenger (e.g. **cyclic AMP**).

Secondary oocytes. Formed from **primary oocytes** *via* the reduction (**meiotic**) division.

Secondary sex characteristics. Characteristics of maleness or femaleness induced by sex **hormone** action (e.g. body hair distribution, voice breaking, antlers).

Secondary spermatocytes. Formed from **primary spermatocytes** *via* the reduction (**meiotic**) division.

Selection. Used in connection with **ovarian** physiology to refer to the growth and development of one (or more) **preovulatory follicles** at the expense of others.

Sella turcica. The depression in the sphenoid bone of the head within which the **hypophysis** lies in some species.

Semen. The ejaculate; contains **seminal fluid** and **sperm**.

Seminal fluid. The secretions of the **testes** and **accessory glands** containing the **sperm**.

Seminal vesicles. **Accessory glands** found in most male mammals.

Seminiferous tubules. Coiled tubes of the **testis** within which **spermatogenesis** takes place.

Senescence. Old age.

Sensitivity. Validity criterion of an **assay**: the minimum concentration of **hormone** measurable by the method.

Serotonin. 5-Hydroxytryptamine.

Sertoli cells. The support cells within the **spermatogenic** elements within the **seminiferous tubules**.

Set-aside. A scheme in which farm land is removed from cultivation.

Sex chromosomes. The **chromosomes** X and Y, involved in sex determination.

Sex ratio. The number of males: females born.

SHBG. Sex **hormone** binding globulin; the specific **binding protein** in blood for **testosterone** and **oestradiol**.

Short-day breeder. A **seasonal breeding** species that breeds in the autumn.

Short-loop feedback. **Hormone feedback** linking the stimulatory signal back on itself or mechanisms controlling it (e.g. **gonadotrophins** controlling **gonadotrophin** release).

Short luteal phase. A situation in which the **corpus luteum** remains active for a shorter than normal lifespan.

Side-chain cleavage. The removal of all or part of a side chain during **steroid** biosynthesis.

Side-effects. Effects of a drug (**hormone**) (e.g. **contraceptive**) which are extra (usually unwanted) to the main effect.

Silastic implant. A reservoir for a continuous release **hormone** implant.

Specificity. A validity criterion of an **assay**, involving checking for lack of interference by related **hormones** or other substances.

Sperm. The male **gamete**.

Sperm capacitation. See **capacitation**.

Sperm hyperactivity. See **hyperactive sperm**.

Spermatids. **Haploid** cells formed from **secondary spermatocytes** during **spermatogenesis**.

Spermatocytes. Immature male germ cells, see **primary** or **secondary spermatocytes**.

Spermatogenesis. The process of production of male **gametes**.

Spermatogenic cycle. The series of changes taking place at any portion of a **spermatogenic tubule** between two appearances of the same stages of development.

Spermatogenic wave. Spatial, sequential order of stages of **spermatogenesis** along the length of a **seminiferous tubule** at any given time.

Spermatogonia. **Diploid** male germ cells; divide by **mitosis** to increase numbers and differentiate into **primary spermatocytes** during **spermatogenesis**.

Spermatozoon. A **sperm** (plural, spermatozoa).

-Spermia. Pertaining to a **semen** abnormality.

Spermicide. A substance for killing **sperm**.

Spermiogenesis. The process of **spermatid** maturation into **sperm**.

Spina bifida. A congenital defect in the closure of the spinal canal.

Spinnbarkheit test. The formation of **mucus** threads when **cervical mucus** is stretched between glass slides; test of **oestrogen** action (i.e. to estimate time of **oestrus** and **ovulation**).

Sponge. **Barrier contraceptive** in the female.

Spongiform encephalopathy. Spongy brain disease.

Spontaneous luteal phase. Occurs in most mammals in which an active **luteal phase** of the cycle does not require a **mating** stimulus.

Spontaneous ovulator. A mammal that does not require a **mating** stimulus to trigger **ovulation**.

Spotting. A minor **breakthrough bleed**.

Stein–Leventhal syndrome. A specific type of **polycystic ovary** disease in women; symptoms include **anovulation, amenorrhoea, hirsutism** and obesity.

Stem cells. Cells which retain the ability to divide and differentiate into other cell types.

Stem cells of the blastocyst. The cells of the undifferentiated **inner cell mass**.

Stem cells of the testis. Spermatogonia which continue to undergo **mitosis**, from which differentiated **spermatogonia** arise that form **primary spermatocytes**.

Sterile male technique. A method in which **sterile** males are released into a **population** to lower the **breeding** potential of that **population**.

Sterilization. Any operation on either sex to permanently prevent **fertility**, usually by ligating or cutting the **fallopian tube** or **vas deferens**.

Sterility. Irreversible **infertility**.

Steroid biosynthetic pathway. The general biosynthetic pathway for **steroids** involving a sequence from **cholesterol** through 21 carbon **progestogens** to either 21 carbon **androgens** and 18 carbon **oestrogens** or 21 carbon **corticosteroids**.

Steroid glucuronides. Steroid conjugates with glucuronic acid.

Steroid nucleus. The basic 17 carbon, 4 ring structure of all **steroids**.

Steroidogenesis. The biosynthesis of **steroids**.

Steroids. Lipid-soluble **hormones** chemically related to **cholesterol** (e.g. **androgens, oestrogens, progestogens, corticosteroids**).

Steroid sulphates. Steroid conjugates with a sulphuric acid residue.

Streak ovaries. Ovaries devoid of **oocytes** as in **Turner's syndrome**.

Stress. The physiological effect of adverse stimuli.

Stress response. Response of an animal to **stress**, including the increase of secretion of **glucocorticoids**.

Substitution of groups on a molecule. The replacement of one chemical group with another.

Subunits. Components of **peptides** and **proteins**, including the **alpha** and **beta subunits** of the **glycoprotein hormones, inhibins** and **activins**.

Suckling. The process by which offspring obtain milk from the nipple.

Suckling reflex. The initiation of **oxytocin** release and **milk** ejection (let-down) by **suckling**.

Sulphates (steroid). See **steroid sulphates**.

Superovulation. Release of more **oocytes** at ovulation than is characteristic for the species.

Surrogate mother. A substitute mother; the use of a second female to carry the offspring of another after **embryo transfer**.

Swim-up techniques. Methods of separating motile **sperm** by allowing them to 'swim up' into media.

Synapse. The region where one nerve cell makes functional contact with another.

Synchronization of oestrus and/or ovulation. The process whereby a group of animals are induced to come into **oestrus** and **ovulate** at the same time.

Tail. End portion of **sperm**.

Target organ. An organ containing specific **receptors** (or other mechanisms) for responding to a particular **hormone**.

Taurine. Sulphur-containing **amino acid**; functions may include actions on **sperm hyperactivity**.

Teasers. Animals used to aid, monitor or induce **reproduction** (e.g. an **oestrous female** with which a male mounts but does not **mate**, for aiding **semen** collection); see also **vasectomized rams**.

Teasing. The use of a **teaser**.

Teratogen. A substance causing defects in the offspring of a mother exposed to the substance.

Testis. Male **gonad**.

Testosterone. The major **androgen** produced by the **testis**.

Testosterone oenanthate. An ester of **testosterone**.

Testosterone propionate. An ester of **testosterone**.

Tetanus toxoid. Inactivated tetanus toxin; used as a carrier to develop **antibodies** against **hCG**.

TGF. Transforming growth factor.

Theca cells. Cell layers surrounding the **granulosa cells** of **ovarian follicles**.

Theca externa. Theca cells on the outside, comprising mostly connective tissue.

Theca interna. Blood-vessel-carrying, **steroid**-synthesizing **theca cells**, on the inside of **theca externa**.

Threatened species. See **endangered species**.

Three mile island. An island in the Shenandoah river, near Harrisburg, Pennsylvania; famous as the site of a power station which almost suffered a catastrophic accident to a pressurized water reactor on 28th March 1979.

Thyroid glands. Specific **endocrine glands** in the neck.

Thyroid hormones. Hormones from the **thyroid** gland, including **triiodothyronine** and **thyroxine**.

Thyroxine. Secreted by the **thyroid gland**, must be converted to **triiodothyronine** for many actions.

TIFT. Trans**cervical** intra**fallopian** transfer.

Transforming growth factors. α and β; **peptides** which have important

autocrine and **paracrine** actions in the **testis** and **ovary**.

Transgenic. An organism containing an artificially transferred **gene**.

Transferable embryos. Embryos produced by **superovulation** or **IVF** that are considered of good enough quality for **embryo transfer**.

Transferrin. Protein that binds iron, secreted by **Sertoli cells**.

Tricycling system. The taking of three or four packets of **combined oral contraceptives** in succession (i.e. a pill every day for 63 days), before a break during which a **period** normally occurs.

Triiodothryonine. A **thyroid hormone** may be the active form of **thyroxine**.

Trilostane. A 3β-**hydroxysteroid dehydrogenase** inhibitor.

Triphasic pills. Combined oral contraceptives in which the **oestrogen: progestogen** ratio varies in three phases.

Trisomy. A condition in which a **chromosome** is present in triplicate instead of the normal pair.

Tritium. ³H, **radioisotope** of hydrogen used to label many **steroids** for **radioimmunoassay**.

Trophoblast. The outer layer of a mammalian **blastocyst**, which forms extra-embryonic membranes (e.g. **chorion**) and the **embryonic** part of the **placenta**.

Trophoblast proteins. Specific **interferon**-like **proteins** produced by **embryos** in certain species which act as **pregnancy recognition hormones**.

Tubal block. An obstruction of the **fallopian tubes** leading to **infertility**.

TUFT. Transuterine **fallopian transfer**.

Tumour. An abnormal growth, may be either benign or **malignant**.

Turner's syndrome. Associated with females with a missing sex **chromosome** (XO); leads to the possession of **streak ovaries** containing no **oocytes**.

Twinning. The production of two offspring in a normally **monotocous** species; they may be either mono**zygotic** (identical) or di**zygotic** (non-identical); or the process of producing twins.

Two cell: two hormone hypothesis. The action of **LH** on **theca cells** to produce **androgen** which is passed to **granulosa cells** for conversion to **oestrogen** under **FSH** control.

Ultrasonography. Visualization of structures using **ultrasound**.

Ultrasound. Sound waves of high frequency used in devices to visualize internal structures (**follicles, embryos**) by **ultrasonography**.

Ultrashort-loop feedback. Hormone feedback linking for example the **hypothalamic** secretions with their own control (e.g. GnRH controlling **GnRH** release); alternative name for **autocrine** control.

Ungulates. Hoofed mammals.

Urbanization. The movement of human **population** from the country to the town.

Urethral glands. **Accessory glands** in the males of some species.

Useable embryos. See **transferable embryos**.

Uterus. The organ within which the **fetus** develops.

Vacuum aspiration. Method of inducing **abortion** by suction.

Vagina. Links the **cervix** to the **vulva**, receives the **sperm** at **coitus** and serves as the **birth** canal.

Vaginal ring. A **hormone**-releasing device for the **vagina**, used to release **contraceptive** amounts of **steroid**.

Vaginal sponges. Usually **progestogen**-containing devices used for **oestrous synchronization**, particularly in sheep.

Varicocele. Dilation of the peritesticular veins of the **testis**; may lead to **infertility**.

Vas deferens. The tube in the male for **sperm** transport from the **epididymis** to the base of the **penis**.

Vasectomized rams. Used as **teasers** to induce early **seasonal** onset of reproductive activity in ewes by means of the **ram effect**.

Vasectomy. Male **sterilization** by cutting or blocking the **vas deferens**.

Ventral prostate test. A **bioassay** for the measurement of **LH** involving studying the increase in weight of the ventral **prostate** in immature **hypophysectomized** rats.

Vesicular follicle. An antral **follicle** (one with an **antrum**).

Virilization. Masculinization; also acquisition of male characteristics by the female.

Vitamin A. A fat-soluble (lipid) vitamin; its deficiency affects all tissues.

Vitelline membrane. The thin membrane surrounding the **oocyte**.

Vitrification. Osmotic dehydration prior to cooling by controlled equilibration in a highly concentrated solution of **cryoprotective agent**.

Viviparity. The bearing of live young, nurtured within the body of the mother.

Vomeronasal organ. Sensory structure in or near the nasal septum, implicated in relaying **pheromonal** signals.

Vulva. The female external genital structures.

Weaning. Transfer of offspring from mother's milk to other food.

Withdrawal bleed. A bleed induced by removal of the **progestogen** maintenance of the **endometrium** in higher primates.

X chromosome. A sex **chromosome**; normal female mammals are XX, normal males are XY.

X-rays. Electromagnetic rays of very short wave length which can penetrate matter opaque to light rays.

Y chromosome. A sex **chromosome**; normal male mammals carry one Y **chromosome** and one X **chromosome**.

Zero population growth. **Births** and deaths, immigration and emigration are in balance: the **population** remains static.

ZIFT. **Zygote** intra**fallopian transfer**.

Zinc. A high concentration is present in the **prostate** secretion in many species.

Zona-free hamster eggs. Used in tests to assess the fertilizing capacity of **sperm**.

Zona pellucida. The transparent membrane surrounding **oocytes**.

Zona reaction. The change taking place in the **zona** at **fertilization** which prevents **zona** penetration by more than one **sperm**.

-Zoospermia. Pertaining to a **sperm** abnormality.

Zygote. The product of the union of a **sperm** and an **ovum**.

Furthur reading

Chapter 1

Alderson, L. (ed) (1990) *Genetic Conservation of Domestic Livestock*, CAB International, Oxford.

Bennett, P.M. (1990) Establishing breeding programmes for threatened species betweeen zoos. *J. Zool.*, **220**, 513–515.

Blaxter, K.L. (1987) Future hunger. *Lancet*, **1**, 309–313.

Brown, L. (1991) *The State of the World 1990*. Worldwatch/WW Norton, NY.

Cohen, M.N. (1977) *Population Pressure and the Origins of Agriculture*. Yale University Press, New Haven.

Flowerdew, J.R. (1987) *Mammals: Their Reproductive Biology and Population Ecology*. Arnold, London.

Fowler, C. and Smith, T. (1981) *Dynamics of Large Mammal Populations*. Wiley, NY.

Greep, R.O. (1988) Association of birth rates with child health and human welfare: A global view. *Ann. N.Y. Acad. Sci.*, **549**, 166–179.

Hare, R.M. (1988) Possible people. *Bioethics*, **2**, 279–293.

Hutchins, M. and Wiese, R.J. (1991) Beyond genetic and demographic management: the future of the Species Survival Plan and related AAZPA conservation efforts. *Zoo Biol.*, **10**, 285–292.

Keyfitz, N. (1989) The growing human population. *Sci. Am.*, **261**, 118–127.

King, M. and Thornton, J. (1992) Expanding human populations and their ecosystems, in *Ethics in Reproductive Medicine* (eds D.R. Bromham, M.E. Dalton, J.C. Jackson and P.J.R. Millican) Springer-Verlag, London, pp. 59–69.

Norton, B.G. (1988) *Why Preserve Natural Variety*. Princeton University Press.

Peel, L., Tribe, D.E. (eds) (1983) *Domestication, Conservation and Use of Animal Resources*. Elsevier, Amsterdam.

Phillips, R.W. (1980) The population explosion and the future of animal agriculture, in *Animal Agriculture* 2nd edn. (eds H.H. Cole and W.N. Garrett) Freeman, NY, pp. 46–57.

Pinder, N.J. and Barkham, J.P. (1978) An assessment of the contribution of captive breeding to the conservation of rare mammals. *Biol. Conserv.*, **13**, 187–245.

Norton, B.G. (1988) *Why Preserve Natural Variety*. Princeton University Press.

Robinson, J.J. (1990) The pastoral animal industry in the 21st century. *Proc. NZ Soc. Anim. Prod.*, **50**, 345–359.

Setchell, B.P. (1992) Domestication and reproduction. *Anim. Reprod. Sci.*, **28**, 195–202.

Smith, G.R. and Hearn, J.P. (eds) (1988) *Reproduction and Disease in Captive and Wild Animals. Symp. Zool. Soc. Lond No 60*. Clarendon Press, Oxford.

UNFPA (United Nations Population Fund) (1991) *The State of World Population*.

Wildt, D.E. (1989) Reproductive research in conservation biology: Priorities and avenues for support. *J. Zoo Wildl. Med.*, **20**, 391–395.

Chapter 2

Adashi, E.Y., Resnick, C.E., D'Ercole, A.J., *et al.* (1985) Insulin-like growth factors as intraovarian regulators of granulosa cell growth and function. *Endocrine Rev.*, **6**, 400–420.

Aitken, J. (1991) Do sperm find eggs attractive? *Nature*, **35**, 19–20.

Asch, R.H., Balmaceda, J.P. and Johnston, I. (eds) (1990) *Gamete Physiology*. Serono Symposia, Norwell.

Austin, C.R. and Short, R.V. (eds) (1982) *Germ cells and Fertilization. Reproduction in Mammals, Book 1*, 2nd Edn. Cambridge University Press.

Austin, C.R. and Short, R.V. (eds) (1984) *Hormonal Control of Reproduction. Reproduction in Mammals, Book 3*, 2nd Edn. Cambridge University Press.

Bagnell, C.E. (1991) Production and biologic action of relaxin within the ovarian follicle: an overview. *Steroids*, **56**, 242–246.

Baird, D.T. (1987) A model for follicular selection and ovulation: Lessons from superovulation. *J. Steroid Biochem.*, **27**, 15–23.

Baird, D.T., Campbell, B.K., Mann, G.E. and McNeilly, A.S. (1991) Inhibin and oestradiol in the control of FSH secretion in the sheep. *J. Reprod. Fertil.*, **Suppl. 43**, 125–138.

Barb, C.R., Kraeling, R.R. and Rampacek, G.B. (1991) Opioid modulation of gonadotropin and prolactin secretion in domestic animals. *Domestic Anim. Endocrinol.*, **8**, 15–27.

Bavister, B.D., Cummins, J. and Roldan, E.R.S. (eds) (1990) *Fertilization in Mammals*. Serono Symposia, Norwell.

Bazer, F.W. (1992) Mediators of maternal recognition of pregnancy in mammals. *Proc. Soc. Exp. Biol. Med.*, **199**, 373–384

Bazer, F.W., Thatcher, W.W., Hansen, P.J., *et al.* (1991) Physiological mechanisms of pregnancy recognition in ruminants. *J. Reprod. Fertil.*, **Suppl. 43**, 39–47.

Braden, T.D. and Conn, P.M. (1991) GnRH and its actions. *J. Physiol. Pharmacol.*, **69**, 445–458.

Brooks, N., Challis, J., McNeilly, A. and Doberska, C. (eds) (1992) *Frontiers in Reproductive Biology. J. Reprod. Fertil.*, **Suppl. 45**.

Burger, H. and de Kretser, D. (eds) (1989) *The Testis*, 2nd Edn. Raven Press, NY.

Burks, D.J. and Saling, P.M. (1992) Molecular mechanisms of fertilization and activation of development. *Anim. Reprod. Sci.*, **28**, 79–86.

Challis, J.R.G. and Lye, S.J. (1986) Parturition. *Oxf. Rev. Reprod. Biol.*, **8**, 61–129.

Chard, T. (1991) Interferon-α is a reproductive hormone. *J. Endocrinol.*, **131**, 337–338.

Cole, D.J.A., Foxcroft, G.R. and Weir, B.J. (eds) (1990) *Control of Pig Reproduction III. J Reprod. Fertil.*, **Suppl. 40**.

Conn, P.M. and Crowley, W.F. (1991) GnRH and its analogues. *New Engl. J. Med.*, **324**, 93–103.

Cupps, P.T. (ed) (1991) *Reproduction in Domestic Animals*, 4th Edn. Academic Press, California.

Dalton, M. (1989) *Gynaecological Endocrinology*. MacMillan, London.

De Kretser, D.M. (ed) (1992) The Testes. *Baillière's Clin. Endocrinol. Metab.*, **6(2)**.

De Kretser, D.M., Sun, Y.T., Drummond, A.E., *et al.* (1990) Physiological mechanisms controlling spermatogenesis, in *Gamete Physiology* (eds R.H. Asch, J.P. Balmaceda and I. Johnston) Serono Symposia, Norwell, pp. 19–29.

Desjardins, C. and Ewing, L. (eds) (1991) *Cell and Molecular Biology of the Testis.* Oxford University Press.

Dobson, H. (1988) Softening and dilation of the cervix. *Oxf. Rev. Reprod. Biol.,* **10**, 491–514.

Draincourt, M.A. (1991) Follicular dynamics in sheep and cattle. *Theriogenology,* **35**, 56–79.

Drobnis, E.Z. and Overstreet, J.W. (1992) Natural history of mammalian spermatozoa in the female reproductive tract. *Oxf. Rev. Reprod. Biol.,* **14**, 1–45.

Dunbar, B.S. and Orand, M.G. (eds) (1991) *A Comparative Overview of Mammalian Fertilization.* Plenum Publishing, NY.

Edwards, R.G. (1980) *Conception in the Human Female.* Academic Press, London.

Findlay, J.K. (1991) The ovary. *Baillière's Clin. Endocrinol. Metab.,* **5**, 755–769.

Findlay, J.K., Robertson, D.M., Clarke, I.J., *et al.* (1992) Hormonal regulation of reproduction – general concepts. *Anim. Reprod. Sci.,* **28**, 319–328.

Findlay, J.K. and Salamonsen, L.A. (1991) Paracrine regulation of implantation and uterine function. *Baillière's Clin. Obstet. Gynaecol.,* **5**, 117–132.

Flint, A.P.F., Hearn, J.P. and Michael, A.E. (1990) The maternal recognition of pregnancy in mammals. *J. Zool.,* **221**, 327–341.

Fortune, J.E., Sirois, J., Turzillo, A.M. and Lavoir, M. (1991) Follicle selection in domestic ruminants. *J. Reprod. Fertil.,* **Suppl. 43**, 187–198.

Frandson, R.D. (1981) *Anatomy and Physiology of Farm Animals,* 3rd Edn. Lea and Febiger, Philadelphia, Chap. 25–29.

Fritz, I.B. (1990) Cell–cell interactions in the testis: a guide for the perplexed, in *Biology of Mammalian Germ Cell Mutagenesis* (eds J.W. Allen, B.A. Bridges, M.F. Lyon *et al.*) Cold Spring Harbor Laboratory Press, NY, pp. 19–34.

Gandolfi, F., Brevini, T.A.L., Modina, S. and Passoni, L. (1992) Early embryonic signals: embryo–maternal interactions before implantation. *Anim. Reprod. Sci.,* **28**, 269–276.

Geisert, R.D., Short, E.C. and Zavy, M.T. (1992) Maternal recognition of pregnancy. *Anim. Reprod. Sci.,* **28**, 287–298.

Goto, K. and Iritani, A. (1992) Oocyte maturation and fertilization. *Anim. Reprod. Sci.,* **28**, 407–413.

Gower, D.B. (1979) *Steroid Hormones.* Croom Helm, London.

Greep, R.O., Koblinsky, M.A. and Jaffe, F.S. (eds) (1976) *Reproduction and Human Welfare; A Challenge to Research.* MIT Press, MA.

Greer, I.A. (1991) Physiological control of parturition. *Contemp. Rev. Obstet. Gynecol.,* **3**, 209–218.

Hafez, E.S.E. (ed) (1991) *Assisted Human Reproductive Technology. Reproductive Health Technology.* Hemisphere Publ. Co., NY.

Hamilton, D.W. and Waites, G.M.H. (eds) (1990) *Cellular and Molecular Events in Spermiogenesis.* Cambridge University Press.

Hartshorne, G.M. and Edwards, R.G. (1991) Role of embryonic factors in implantation: recent developments. *Baillière's Clin. Obstet. Gynaecol.,* **5**, 133–158.

Haynes, N.B., Lamming, G.E., Yang, K.-P., Brooks, A.N. and Finnie, A.D. (1989) Endogenous opioid peptides and farm animal reproduction. *Oxf. Rev. Reprod. Biol.,* **11**, 111–145.

Hillier, S.G. (ed) (1991) *Ovarian Endocrinology.* Blackwell, London.

Hillier, S.G. (1991) Regulatory functions for inhibin and activin in human ovaries. *J. Endocrinol.*, **131**, 171–175.

Hillier, S.G. (1990) Ovarian manipulation with pure gonadotrophins. *J. Endocrinol.*, **127**, 1–4.

Hillier, S.G. (1991) Paracrine control of follicular estrogen synthesis. *Sem. Reprod. Endocrinol.*, **9**, 332–340.

Hirshfield, A.N. (ed) (1989) *Growth Factors and the Ovary*. Plenum Publishing Co, NY.

Hochberg, R.B. and Naftolin, F. (1991) *The New Biology of Steroid Hormones*. Serono Symposia, NY.

Hunter, M.G. and Wiesak, T. (1990) Evidence for and implications of follicular heterogeneity in pigs. *J. Reprod. Fertil.*, **Suppl. 40**, 163–177.

Hunter, R.H.F. (1980) *Physiology and Technology of Reproduction in Female Domestic Animals*. Academic Press, London.

Hunter, R.H.F. (1982) *Reproduction in Farm Animals*. Longman, London.

Ireland, J.J. (1987) Control of follicular growth and development. *J. Reprod. Fertil.*, **Suppl. 34**, 39–54.

Isidori, A., Fabbri, A. and Dufau, M.L. (eds) (1990) *Normal Communicating Events in the Testis*. Raven Press, NY.

Jeffcoate, S.L. and Hutchinson, J.S.M. (eds) (1978) *The Endocrine Hypothalamus*. Academic Press, London.

Johnson, L. (1990) Spermatogenesis, in *Gamete Physiology* (eds R.H. Asch, J.P. Balmaceda and I. Johnston) Serono Symposia, Norwell, pp. 3–18.

Johnson, M.H. and Everitt, B.J. (1988) *Essential Reproduction*, 3rd Edn. Blackwell, London.

Jones, R. (1990) Identification and functions of mammalian sperm–egg recognition molecules during fertilization. *J. Reprod. Fertil.*, **Suppl.. 42**, 89–105.

Jones, R.E. (1991) *Human Reproductive Biology*. Academic Press, San Diego.

Jones, T.H., Brown, B.L. and Dobson, P.R.M. (1990) Paracrine control of anterior pituitary hormone secretion. *J. Endocrinol.*, **127**, 5–13.

Karsch, F.J., Moenter, S.M. and Caraty, A. (1992) The neuroendocrine signal for ovulation. *Anim. Reprod. Sci.*, **28**, 329–341.

Knobil, E. (1980) The neuroendocrine control of the menstrual cycle. *Rec. Progr. Horm. Res.*, **36**, 53–88.

Knobil, E. and Neill, J.D. (eds) (1988) *The Physiology of Reproduction*. Raven Press, NY.

Krey, L.C., Gulyas, B.J. and McCracken, J.A. (eds) (1989) *Autocrine and Paracrine Mechanisms in Reproductive Endocrinology*. Plenum Publishing, NY.

LaBarbera, A.R. and Rebar, R.W. (1990) Reproductive peptide hormones: generation, degradation, reception and action. *Clin. Obstet. Gynecol.*, **33**, 576–590.

Lamming, G.E. (ed) (1990) *Marshall's Physiology of Reproduction, Vol. 2. The Male*. 4th Edn., Churchill Livingstone, Edinburgh.

Lamming, G.E., Flint, A.P.F. and Weir, B.J. (eds) (1991) Reproduction in Domestic Ruminants II. *J. Reprod. Fertil.*, **Suppl. 43**.

Lincoln, D.W., McNeilly, A.S. and Sharpe, R.M. (1989) Reproductive physiology of inhibin and related peptides. *Rec. Adv. Endocrinol.* **3**, 77–107.

Lindsay, D.R. and Pearce, D.T. (eds) (1984) *Reproduction in Sheep*. Cambridge University Press.

Lobb, D.K. and Dorrington, J. (1992) Intraovarian regulation of follicular development. *Anim. Reprod. Sci.*, **28**, 343–354.

Maddocks, S. and Setchell, B.P. (1985) The physiology of the endocrine testes. *Oxf. Rev. Reprod. Biol.*, **10**, 53–123.

Marshall, J.C., Dalkin, A.C., Haisenleder, D.J., *et al.* (1991) GnRH pulses, regulators of gonadotropin synthesis and ovulatory cycles. *Rec. Progr. Horm. Res.*, **47**, 155–189.

Mastroianni, L. and Coutifaris, C. (eds) (1991) *The F.I.G.O. Manual of Human Reproduction. Vol. 1, Reproductive Physiology.* Parthenon, Carnforth.

McDonald, L.E. and Pineda, M.H. (eds) (1989) *Veterinary Endocrinology and Reproduction*, 4th Edn. Lea and Febiger, Philadelphia.

Moore, H.D.M. (1990) Development of sperm–egg recognition processes in mammals. *J. Reprod. Fertil.*, **Suppl. 42**, 71–78.

Motta, M. (ed) (1991) *Brain Endocrinology*, 2nd Edn. Raven Press, NY.

Mullaney, B.P. and Skinner, M.K. (1991) Growth factors as mediators of testicular cell–cell interactions. *Baillière's Clin. Endocrinol. Metab.*, **5**, 771–790.

Niswender, G.D., Baird, D.T., Findlay, J.K. and Weir, B.J. (eds) (1987) *Reproduction in Domestic Ruminants. J. Reprod. Fertil.*, **Suppl. 34**.

Peters, A.R. and Ball, P.J.H. (1987) *Reproduction in Cattle*. Butterworths, London.

Roberts, L. (1991) Does the egg beckon sperm when the time is right? *Science*, **252**, 214.

Roberts, R.M. (1989) Conceptus interferons and maternal recognition of pregnancy. *Biol. Reprod.*, **40**, 449–452.

Roberts, R.M. (1991) A role for interferon in early pregnancy. *BioEssays*, **13**, 121–126.

Roberts, R.M., Farin, C.E. and Cross, J.C. (1990) Trophoblast proteins and maternal recognition of pregnancy. *Oxf. Rev. Reprod. Biol.*, **12**, 147–180.

Roche, J.E. and Boland, M.P. (1991) Turnover of dominant follicles in cattle of different reproductive states. *Theriogenology*, **35**, 81–90.

Russel, L.D., Hikim, A.P.S., Ettlin, R.A. and Clegg, E.D. (1990) Mammalian spermatogenesis, in *Histological and Histopathological Evaluation of the Testis* (eds L.D. Russel, R.A. Ettlin, A.P.S. Hikim and E.D. Clegg) Cache River Press, US, pp. 1–40.

Scaramuzzi, R.J., Lincoln, D.W. and Weir, B.J. (eds) (1981) *Reproductive Endocrinology of Domestic Ruminants. J. Reprod. Fertil.*, **Suppl. 30**.

Schomberg, D.W. (ed) (1991) *Growth Factors in Reproduction*. Springer-Verlag, NY.

Sharpe, R.M. (1990) Review. Intratesticular control of steroidogenesis. *Clin Endocrinol.*, **33**, 787–807.

Shemesh, M. and Weir, B.J. (eds) (1989) *Maternal Recognition of Pregnancy and Maintenance of the Corpus Luteum. J. Reprod. Fertil.*, **Suppl. 37**.

Sidhu, K.S. and Guraya, S.S. (1989) Cellular and molecular biology of capacitation and acrosome reaction in mammalian spermatozoa. *Int. Rev. Cytol.*, **118**, 231–280.

Silman, R. (1991) Melatonin and the human gonadotrophin-releasing hormone pulse generator. *J. Endocrinol.*, **128**, 7–11.

Simon, A., Birkenfeld, A. and Schenker, J.G. (1990) GnRH: mode of action and clinical applications. A review. *Int. J. Fertil.*, **35**, 350–362.

Sizonenko, P.C. and Aubert, M.L. (eds) (1990) *Developmental Endocrinology*. Raven Press, NY.

Skinner, M.K. (1991) Cell–cell interactions in the testis. *Endocrinol. Rev.*, **12**, 45–77.

Smith, S.K. (1991) The role of prostaglandins in implantation. *Baillière's Clin. Obstet. Gynaecol.*, **5**, 73–94.

Taverne, M.A.M. (1992) Physiology of parturition. *Ann. Reprod. Sci.*, **28**, 433–440.

Taya, K., Kaneko, H., Watanabe, G. and Sasamoto, S. (1991) Inhibin and secretion of FSH in oestrous cycles of cows and pigs. *J. Reprod. Fertil.,* **Suppl. 43**, 151–162.

Terranova, P.F. (1991) Regulation of the granulosa cell: growth factor interactions. *Sem. Reprod. Endocrinol.,* **9**, 313–320.

Thiéry, J.C. and Martin, J.B. (1991) Neuro-physiological control of the secretion of gonadotrophin-releasing hormone and luteinizing hormone in the sheep: a review. *Reprod. Fertil. Dev.,* **3**, 137–173.

Thorburn, G.D. (1991) The placenta, prostaglandins and parturition. A review. *Reprod. Fertil. Dev.,* **3**, 277–294.

Thorburn, G.D., Holingworth, S.A. and Hooper, S.B. (1991) The trigger for parturition in sheep: fetal hypothalamus or placenta. *J. Dev. Physiol.,* **15**, 71–79.

Tonetta, S.A. and DiZerega, G.S. (1989) Intragonadal regulation of follicular maturation. *Endocrine Rev.,* **10**, 205–229.

Tyndale-Biscoe, H. and Renfree, M. (1987) *Reproductive Physiology of Marsupials.* Cambridge University Press.

Urban, R.J. and Veldhuis, J.D. (1992) Endocrine control of steroidogenesis in granulosa cells. *Oxf. Rev. Reprod. Biol.,* **14**, 225–262.

Veldhuis, J.D. (1990) The hypothalamic pulse generator: the reproductive core. *Clin. Obstet. Gynecol.,* **33**, 538–550.

Wasserman, P.M. (ed) (1990) *Elements of Mammalian Fertilization Volumes 1 and 2.* Wolfe Publishing, London.

Weir, B.J. and Rowlands, I.W. (1973) Reproductive strategies of mammals. *Annu. Rev. Ecol. Syst.,* **4**, 139–163.

Witt, B.R. and Thorneycroft, I.H. (1990) Reproductive steroid hormones: generation, degradation, reception and action. *Clin. Obstet. Gynecol.,* **33**, 563–575.

Yen, S.S.C. (1990) Clinical endocrinology of reproduction. in *Hormones: From Molecules to Disease* (eds E.-E. Baulieu and P.A. Kelly) Hermann, Paris, pp. 445–450.

Yen, S.S.C. and Jaffe, R.B. (1986) *Reproductive Endocrinology,* 2nd Edn. Saunders, Philadelphia.

Yen, S.S.C. and Vale, W.W. (eds) (1990) *Neuroendocrine Regulation of Reproduction.* Serono Symposia, Norwell.

Ying, S.-Y. (1988) Inhibins, activins and follistatins: Gonadal proteins modulating the secretion of follicle-stimulating hormone. *Endocrine Rev.,* **9**, 267–293.

Zuckerman, S. and Weir, B.J. (eds) (1979) *The Ovary,* 2nd Edn. Academic Press, NY.

Chapter 3

Aimen, J. (ed) (1984) *Infertility, Diagnosis and Management.* Springer Verlag, NY.

Al-Azzawi, F. (1992) Endocrinological aspects of the menopause. *Br. Med. Bull.,* **48**, 262–275.

Anon (1991) Declining fertility: egg or uterus. *Lancet,* **338**, 285–286.

Anon (1991) Pathology of Reproductive Infertility. *Monogr. Pathol.,* **33**, 1–348.

Arendt, J. (1986) Role of the pineal gland and melatonin in seasonal reproductive function in mammals. *Oxf. Rev. Reprod. Biol.,* **8**, 266–320.

Austin, C.R. and Short, R.V. (eds) (1984) *Reproductive Fitness. Reproduction in Mammals, Book 4,* 2nd Edn. Cambridge University Press.

Barnes, R.B. (1991) Polycystic ovary syndrome and ovarian steroidogensis. *Sem. Reprod. Endocrinol.,* **9**, 360–366.

Betteridge, K.J. (1991) Current concepts of the importance and etiology of embryonic death in farm animals. *Revt. Bras. Reprod. Anim.*, **3 (Suppl. 1)**, 66–73.

Booth, W.D. and Signoret, J.P. (1992) Olfaction and reproduction in ungulates. *Oxf. Rev. Reprod. Biol.*, **14**, 263–301.

Bronson, F.H. (1989) *Mammalian Reproductive Biology*. Univ. Chicago Press.

Carr, B.R. and Griffin, J.E. (1985) Fertility and its complications. in *William's Textbook of Endocrinology*, 7th Edn (eds J.D. Wilson and D.W. Foster) Saunders, Philadelphia, pp. 452–475.

Christian, J.J. (1980) Endocrine factors in population regulation. in *Biosocial Mechanisms in Population Regulation* (eds M.N. Cohen, R.S. Malpass and H.G. Klein) Yale University Press, New Haven, pp. 55–115.

Clark, I.J. and Tilbrook, A.J. (1992) Influence of non-photoperiodic environmental factors on reproduction in domestic animals. *Anim. Reprod. Sci.*, **28**, 219–228.

Cohen, I. and Speroff, L. (1991) Premature ovarian failure: an update. *Obstet. Gynecol. Surv.*, **46**, 156–162.

Colpi, G.M. and Pozza, D. (eds) (1992) *Diagnosing Male Infertility. New Possibilities and Limits. Progr. Reprod. Biol. Med. 15*. Karger, Basel.

De Cherney, A.H. (ed) (1986) *Reproductive Failure*. Churchill Livingstone, Edinburgh.

Desouza, M.J. and Metzger, D.A. (1991) Reproductive dysfunction in amenorrheic athletes and anorexic patients: A review. *Med. Sci. Sport Exercise*, **23**, 995–1007.

Dunaif, A., Givens, J.R., Haseltine, F.P. and Merriam, G.R. (eds) (1992) *Polycystic Ovary Syndrome. Current Issues in Endocrinology and Metabolism*. Blackwells, Oxford.

Eddy, R.G. and Ducker, M.J. (eds) (1984) *Dairy Cow Fertility*. Bull. Vet. Assoc. Editorial Services, London.

Galina, C.S. and Arthur, G.H. (1989) Review of cattle reproduction in the tropics 3. Puerperium. *Anim. Breed. Abs.*, **57**, 899–910.

Galina, C.S. and Arthur, G.H. (1990) Review of cattle reproduction in the tropics 4. Oestrous cycles. *Anim. Breed. Abs.*, **58**, 697–707.

Galina, C.S. and Arthur, G.H. (1991) Review of cattle reproduction in the tropics 6. The male. *Anim. Breed. Abs.*, **59**, 403–412.

Garverick, H.A., Zollers, W.G. and Smith, M.F. (1992) Mechanisms associated with corpus luteum lifespan in animals having normal and subnormal luteal function. *Anim. Reprod. Sci.*, **28**, 111–124.

Gilmore, D. and Cook, B. (eds) (1981) *Environmental Factors in Mammalian Reproduction*. MacMillan, London.

Greenhall, E. and Vessey, M. (1990) The prevalence of subfertility: a review of the current confusion and a report on two new studies. *Fertil. Steril.*, **54**, 978–983.

Harcourt, A.H. (1987) Dominance and fertility among female primates. *J. Zool.*, **213**, 471–487.

Hargreaves, T.B. (ed) (1983) *Male Infertility*. Springer-Verlag, Berlin.

Harrison, R.F. (1991) Aims and objectives in the infertility clinic: the practical issues. *Int. J. Fertil.*, **36**, 204–211.

Hunter, M.G. (1991) Characteristics and causes of the inadequate corpus luteum. *J. Reprod. Fertil.*, **Suppl. 43**, 91–99.

Hussain, A.M. and Daniel, R.C.W. (1991) Bovine normal and abnormal reproductive and endocrine functions during the postpartum period: A review. *Reprod. Domest. Anim.*, **26**, 101–111.

Hyland, J.H. (1990) Review: reproductive endocrinology, the role in fertility and infertility in the horse. *Br. Vet. J.*, **146**, 1–16.

I'Anson, H., Foster, D.L., Foxcroft, G.R. and Booth, P.J. (1991) Nutrition and reproduction. *Oxf. Rev. Reprod. Biol.*, **13**, 239–311.

Jaffe, S.B. and Jewelewicz, R. (1991) The basic infertility investigation. *Fertil. Steril.*, **56**, 599–613.

Korenman, S.G. (ed) (1990) *The Menopause*. Serono Symposia, Norwell.

Kraus, F.J. and Damjanov, I. (eds) (1991) *Pathology of Reproductive Failure*. Williams and Wilkins, Baltimore.

Laing, J.A. (ed) (1979) *Fertility and Infertility in Domestic Animals*, 3rd Edn. Baillière Tindall, London.

Lamming, G.E. (ed) (1984) *Marshall's Physiology of Reproduction*. 4th Edn. Vol. 1 *Reproductive Cycles of Vertebrates*. Churchill Livingstone, Edinburgh.

Lauderdale, J.W. (1986) A review of patterns of change in luteal function. *J. Anim. Sci.*, **62 (Suppl. 2)**, 79–91.

Lincoln, D.W. (1989) Reproductive 'clocks'. *Res. Reprod. Wall Charts*.

Lincoln, G.A. (1992) Photoperiod-pineal-hypothalamic relay in sheep. *Anim. Reprod. Sci.*, **28**, 203–217.

Lincoln, G.A. and Short, R.V. (1980) Seasonal breeding: nature's contraceptive. *Rec. Progr. Horm. Res.*, **36**, 1–43.

Marchlewska-Koj, A. (1984) Pheromones and mammalian reproduction. *Oxf. Rev. Reprod. Biol.*, **6**, 266–302.

Markham, S. (1991) Cervico-utero-tubal factors in infertility. *Curr. Opin. Obstet. Gynecol.*, **3**, 191–196.

Martin, G.B., Oldham, C.M., Gognié, Y. and Pearce, D.T. (1986) The physiological responses of anovulatory ewes to the introduction of rams: a review. *Livestock Prod. Sci.*, **15**, 219–247.

McEntee, K. (1990) *Reproductive Pathology of Domestic Mammals*. Academic Press, NY.

Milligan, S.R. (1982) Induced ovulation in mammals. *Oxf. Rev. Reprod. Biol.*, **4**, 1–46.

Moberg, G.P. (1991) How behavioural stress disrupts the endocrine control of reproduction in domestic animals. *J. Dairy Sci.*, **74**, 304–311.

Oak, M.K., Vaughan Williams, C.A. and Elstein, M. (1983) The current status of infertility associated with endometriosis. *J. Clin. Reprod. Infertil.*, **2**, 97–112.

O'Callaghan, D., Karsch, F.J. and Roche, J.F. (1992) Melatonin in ewes – a timekeeping hormone regulating seasonal reproductive transitions. *Agric. Biotechnol. News. Inf.*, **4**, 101N–106N.

Ojeda, S.R. (1991) The mystery of mammalian puberty. How much more do we know? *Perspec. Biol. Med.*, **34**, 365–386.

Ortavant, R., Pelletier, J., Ravault, J.P., *et al.* (1985) Photoperiod: main proximal and distal factor of the circannual cycle of reproduction in farm mammals. *Oxf. Rev. Reprod. Biol.*, **7**, 305–345.

Pepperell, R.J., Hudson, B. and Wood, C. (1980) *The Infertile Couple*. Churchill Livingstone, Edinburgh.

Peters, A.R. and Lamming, G.E. (1990) Lactational anoestrus in farm animals. *Oxf. Rev. Reprod. Biol.*, **12**, 245–288.

Reiter, R.J. (1980) Seasonal reproduction: An expedient and essential artifice. *Progr. Reprod. Biol.*, **5**, 1–4.

Renfree, M.B. and Calaby, J.H. (1981) Background to delayed implantation and embryonic diapause. *J. Reprod. Fertil.*, **Suppl. 29**, 1–9.

Robinson, J.J. (1990) Nutrition and the reproduction of farm animals. *Nutr. Res.*

Rev., **3**, 253–276.

Sadlier, R.M.F.S. (1969) *The Ecology of Reproduction in Wild and Domestic Mammals*. Methuen, London.

Sanfilippo, J.S. (1991) Onset of puberty. The advent of a new and intricate level of understanding. *Adolesc. Pediatr. Gynecol.*, **4**, 1–2.

Schlaff, W.D. and Wierman, M.E. (1990) Endocrinology of male fertility and infertility. *Curr. Opin. Obstet. Gynecol.*, **2**, 412–417.

Schwartz, L.B. and Diamond, M.P. (1991) Formation, reduction and treatment of adhesive disease. *Sem. Reprod. Endocrinol.*, **9**, 89–99.

Smith, L.L. (1991) *Alternatives to Infertility*. Raven Press, NY.

Södersten, P. (1989) Hormonal and behavioural rhythms related to reproduction. *Adv. Comp. Environ. Physiol.*, **3**, 1–29.

Spark, R.F.C. (1988) *The Infertile Male: The Clinician's Guide to Diagnosis and Treatment*. Plenum, NY.

Spinage, C.A. (1973) The role of photoperiodism in the seasonal breeding of tropical ungulates. *Mammal Rev.*, **3**, 71–84.

Stewart, D.E., Robinson, G.E., Goldbloom, D.S. and Wright, C. (1990) Infertility and eating disorders. *Am. J. Obstet. Gynecol.*, **163**, 1196–1199.

Tan, S.L. and Jacobs, H.S. (1991) *Infertility: Your Questions Answered*. McGraw Hill, London.

Taymor, M.L. (1990) *Infertility: A Clinician's Guide to Diagnosis and Treatment*. Plenum, NY.

Thatcher, S.S. and Naftolin, F. (1991) The aging and aged ovary. *Sem. Reprod. Endocrinol.*, **9**, 189–199.

Vaughan Williams, C.A. and Elstein, M. (1989) Infertility. *Res. Reprod. Wall Charts*.

Winston, R.M.L. (1986) *Infertility: a Sympathetic Approach*. Martin Dunitz, London.

Wood, J.W. (1989) Fecundity and natural fertility in the human. *Oxf. Rev. Reprod. Biol.*, **11**, 61–109.

Wu, C.H. (1990) Ovulatory disorders and infertility in women with regular menstrual cycles. *Curr. Opin. Obstet. Gynecol.*, **2**, 398–404.

Yovich, J. and Lower, A. (1991) Implantation failure. Clinical Aspects. *Baillière's Clin. Obstet. Gynaecol.*, **5**, 211–252.

Ziegler, T.E. and Bercovitch, F.B. (eds) (1990) *Socioendocrinology of Primate Reproduction. Monographs in Primatology*. Wiley-Liss, NY.

Chapter 4

Alexander, N.J., Griffin, D., Spieler, J.M. and Waites, G.M.H. (eds) (1991) *Gamete Interaction: Prospects for Immunocontraception*. Wiley Liss, NY.

Armstrong, D.T. and Leung, P.C.K. (1990) The physiological basis of superovulation. *Sem. Reprod. Endocrinol.*, **8**, 219–231.

Fletzky, O.A. and Davajan, V. (1990) Management of amenorrhoea and associated disorders. *Curr. Opin. Obstet. Gynecol.*, **2**, 386–391.

Gordon, I. (1983) *Controlled Breeding in Farm Animals*. Pergamon Press, Oxford.

Hamilton, D.W. and Waites, G.M.H. (eds) (1990) *Cellular and Molecular Events in Spermiogenesis*. Cambridge University Press.

McNatty, K.P., Henderson, K.M., Fleming, J.S., *et al.* (1990) How does the F gene influence ovulation rate in booroola ewes. A 1990 perspective. *Proc. NZ Soc. Anim. Prod.*, **50**, 135–140.

Mishell, D.R., Davajan, V. and Lobo, R.A. (eds) (1991) *Infertility, Contraception and Reproductive Endocrinology*, 3rd Edn. Blackwells, Oxford.

Morrow, D.A. (ed) (1980) *Current Therapy in Theriogenology*. Saunders, Philadelphia.

Norman, J. (1991) New drugs: Antiprogesterone. *Br. J. Hosp. Med.*, **45**, 372–375.

Swanson, L.V. (1989) Control of the oestrous cycle in farm animals: A review. *Proc. NZ Soc. Anim. Prod.*, **49**, 71–80.

Taylor, G.T., Griffin, M.G. and Bargett, M. (1991) Search for a male contraceptive. The effect of gossypol on sexual motivation and epididymal sperm. *J. Med.*, **22**, 29–44.

Thatcher, W.W., Hansen, P.J., Plante, C., *et al.* (1990) Understanding and exploiting the physiology and endocrinology of reproduction to enhance reproductive efficiency in cattle. *Proc. NZ Anim. Prod.*, **50**, 109–121.

Tyndale-Biscoe, C.H. (1991) Fertility control in wild-life. *Reprod. Fertil. Dev.*, **3**, 339–343.

Chapter 5

Aitken, R.J., Clarkson, J.S., Huang, G.-F. and Irvine, D.S. (1987) Cell biology of defective sperm function, in *New Horizons in Sperm Cell Research* (ed H. Mohri) Gordon and Breach, NY, pp. 75–89.

Albertson, B.D. (1990) Hormonal assay methodology: Present and future prospects. *Clin. Obstet. Gynecol.*, **33**, 591–610.

Amann, R.P. (1989) Can the fertility of a seminal sample be predicted accurately? *J. Androl.*, **10**, 89–98.

Amann, R.P. (1989) Treatment of sperm to predetermine sex. *Theriogenology*, **31**, 49–60.

Barbieri, R.L. (1991) Radioimmunoassay for reproductive, endocrine and fertility services. *J. Clin. Immunoassay*, **14**, 33–36.

Belsey, M.A., Eliasson, R., Gallegos, A.J., *et al.* (eds) (1980) *Laboratory Manual for the Examination of Human Semen and Semen–Cervical Mucus Interaction*. Press Concern, Singapore.

Berga, S.L. and Daniels, T.L. (1991) Use of the laboratory in disorders of reproductive neuroendocrinology. *J. Clin. Immunoassay*, **14**, 23–28.

Brent, R.L., Jensh, R.P. and Beckman, D.A. (1991) Medical sonography: Reproductive effects and risks. *Teratology*, **44**, 123–146.

Britt, J.H. and Holt, L.C. (1988) Endocrinological screening of embryo donors and embryo transfer recipients: A review. *Theriogenology*, **29**, 189–202.

Butler, J.E. and Biggers, J.D. (1989) Assessing the viability of preimplantation embryos *in vitro*. *Theriogenology*, **31**, 115–126.

Butt, W.R. (ed) (1984) *Practical Immunoassay*. Marcel Dekker, NY.

Chard, T. (1990) *An Introduction to Radioimmunoassay and Related Techniques*, 4th Edn. Elsevier Science, Amsterdam.

Collins, W.P. (1991) The evaluation of reference methods to monitor ovulation. *Am. J. Obstet. Gynecol.*, **165**, 1994–1996.

Cook, B. and Beastall, G.H. (1987) Measurement of steroid hormone concentrations in blood, urine and tissues, in *Steroid Hormones – a practical approach* (eds B. Green and R.E. Leake) IRL Press, Oxford, pp. 1–65.

Crozet, N. (1991) Manipulation of oocytes and *in vitro* fertilization. *J. Reprod. Fertil.*, **Suppl. 43**, 235–243.

Dahl, K.D. and Stone, M.P. (1992) FSH isoforms, radioimunnoassays, bioassays and their significance. *J. Androl.*, **13**, 11–22.

den Daas, N. (1992) Laboratory assessment of semen characteristics. *Anim. Reprod. Sci.*, **28**, 87–94.

Dobson, H. and Davies, D.A.R. (1989) Development of an endocrine challenge test to investigate subfertility in ewes. *Br. Vet. J.*, **145**, 523–530.

Dukelow, W.R. (1980) Captive breeding and laparoscopy in nonhuman primates, in *Current Therapy in Theriogenology* (ed D.A. Morrow) Saunders, Philadelphia, pp. 1142–1150.

Fayez, J.A., Mutie, G. and Schneider, P.J. (1987) The diagnosis value of hysterosalpingography and hysteroscopy in infertility investigation. *Am. J. Obstet. Gynecol.*, **156**, 558–560.

Foote, R.H. (1988) Preservation and fertility prediction of spermatozoa. *11th International Congress on Animal Reproduction and AI*, **5**, 126–134.

Friend, D.R. (1990) Transdermal delivery of contraceptives. *Crit. Rev. Ther. Drug Carrier Syst.*, **7**, 149–186.

Gagnon, C. (ed) (1990) *Controls of Sperm Motility: Biological and Clinical Aspects.* CRC Press, Boca Raton.

Glover, T.D., Baratt, C.L.R., Tyler, T.P.P. and Hennessey, J.F. (1990) *Human Male Fertility and Semen Analysis.* Academic Press, NY.

Gosling, J.P. (1990) A decade of development of immunoassay methodology (review). *Clin. Chem.*, **36**, 1408–1427.

Griffin, P.G. and Ginther, Q.J. (1992) Research applications of ultrasonic imaging in reproductive biology. *J. Anim. Sci.*, **70**, 953–972.

Hall, E.A.H. (1990) *Biosensors.* Oxford University Press.

Hutchinson, J.S.M. (1987) Congenital abnormalities. in *Plastic Surgery in Paediatrics* (ed I.F.K. Muir) Lloyd-Luke, London, pp. 1–20.

Hutchinson, J.S.M. (1988) The interpretation of pituitary gonadotrophin assays – a continuing challenge. *J. Endocrinol.*, **118**, 169–171.

Kähn, W. (1992) Ultrasonography as a diagnostic tool in female animal reproduction. *Anim. Reprod. Sci.*, **28**, 1–10.

Keel, B.A. and Webster, B.W. (eds) (1990) *Handbook of Laboratory Diagnosis and Treatment of Infertility.* CRC Press, Boca Raton.

Leese, H.J. (1992) Metabolism of the preimplantation embryo. *Oxf. Rev. Reprod. Biol.*, **13**, 35–72.

Lehrer, A.R., Lewis, G.S. and Aizinbud, E. (1992) Oestrus detection in cattle: recent developments. *Anim. Reprod. Sci.*, **28**, 355–361.

Lynskey, M.T., Cain, T.P. and Caldwell, B.V. (1991) Laboratory evaluation of the infertile couple and monitoring of gonadotrophin therapy. *J. Clin. Immunoassay*, **14**, 29–32.

May, K. (1991) Home tests to monitor fertility. *Am. J. Obstet. Gynecol.*, **165**, 2000–2006.

McEvoy, J.D. (1992) Alteration of the sex ratio. *Anim. Breed. Abs.*, **60**, 97–111.

McGowan, K.D. and Blakemore, K.J. (1991) Amniocentesis and chorionic villus sampling. *Curr. Opin. Obstet. Gynecol.*, **3**, 221–229.

Monk, M. (1991) Preimplantation diagnosis. *Bibl. Reprod.*, **57**, A1–A8.

Nakamura, R.M. and Stanczyk, F.Z. (1991) Imumunoassays, in *Infertility, Contraception and Reproductive Endocrinology* 3rd Edn. (eds D.R. Mishell, V. Davajan and R.A. Lobo) Blackwell, Oxford, pp. 90–103.

Overstrom, E.W. (1992) Manipulation of early embryonic development. *Anim. Reprod. Sci.*, **28**, 277–285.

Picard, L. and Betteridge, K.J. (1989) The micromanipulation of farm animal embryos, in *Animal Biotechnology: Comprehensive Biotechnology 1st Suppl.* (eds L.A. Babiuk and J.P. Phillips) Pergamon Press, Oxford, pp. 141–175.

Robinson, S., Rodin, D.A., Deacon, A., *et al.* (1992) Which hormone tests for the diagnosis of polycystic ovary syndrome? *Br. J. Obstet. Gynaecol.*, **99**, 232–238.

Rosenfield, R.L. and Helke, J. (1992) Is an immunoassay available for the measurement of bioactive LH in serum? *J. Androl.*, **13**, 1–10.

Russel, L.D., Ettlin, R.A., Hikim, A.P.S. and Clegg, E.D. (1990) *Histological and Histopathological Evaluation of the Testis.* Cache River Press, US.

Schlaff, W.D. (1988) Dynamic testing in reproductive endocrinology, in *Modern Trends in Infertility and Conception Control* (eds E.E. Wallach and R.D. Kempers) Year Book Medical Publishers, Chicago, pp. 149–166.

Senger, P.L. (1991) Electronic approaches. A new future for estrous detection in cattle. *Revta. Bras. Reprod. Anim.*, **3**, 210–220.

Seppala, M., Angervo, M., Koistinen, B., *et al.* (1991) Human endometrial protein secretion relative to implantation. *Baillière's Clin. Obstet. Gynaecol.*, **5**, 61–72.

Tsatsoulis, A., Shalet, S.M. and Robertson, W.R. (1991) Review: bioactive gonadotrophin secretion in man. *Clin. Endocrinol.*, **35**, 193–206.

van der Lende, T., Schasfoort, R.B.M. and van der Meer, R.F. (1992) Monitoring reproduction using immunological techniques. *Anim. Reprod. Sci.*, **28**, 179–185.

van Vliet, R.A., Gibbins, A.M.V. and Walton, J.S. (1989) Livestock embryo sexing: A review of current methods with emphasis on Y-specific DNA probes. *Theriogenology*, **32**, 421–438.

Weitze, K.F. and Petzoldt, R. (1992) Preservation of semen. *Anim. Reprod. Sci.*, **28**, 229–235.

White, K.L. (1989) Embryo and gamete sex selection, in *Animal Biotechnology: Comprehensive Biotechnology 1st Suppl* (eds A. Babiuk and J.P. Phillips) Pergamon Press, Oxford, pp. 179–202.

WHO (1987) *Laboratory Manual for the Examination of Human Semen and Semen– Cervical Mucus Interaction.* Cambridge University Press.

Chapter 6

Allen, W.R. (1991) Extra-species embryo transfer in equids, in *European Embryo Transfer Association 7th Scientific Meeting Cambridge*, INRA, Nouzilly, pp. 87–100.

Anderson, G.B. (1983) Embryo transfer in domestic animals. *Adv. Vet. Sci. Comp. Med.*, **27**, 129–162.

Asch, R.H. (1990) GIFT and associated techniques, in *Gamete Physiology* (eds R.H. Asch, J.P. Balmaceda, I. Johnston) Serono Symposia, Norwell, pp. 287–304.

Austin, C.R. and Short, R.V. (eds) (1986) *Manipulating Reproduction. Reproduction in Mammals, Book 5*, 2nd Edn. Cambridge University Press.

Babiuk, L.A. and Phillips, J.P. (eds) (1989) *Animal Biotechnology: Comprehensive Biotechnology 1st Suppl.* Pergamon Press, Oxford.

Ballou, J.D. (1992) Potential contribution of cryopreserved germ plasm to the preservation of genetic diversity and conservation of endangered species in captivity. *Cryobiology*, **29**, 19–25.

Barratt, C.L.R., Chauhan, M. and Cooke, I.D. (1990) Donor insemination – a look to the future. *Fertil. Steril.*, **54**, 375–387.

Bazer, F.W. and Johnson, H.M. (1991) Type 1 conceptus interferons: maternal recognition of pregnancy signals and potential therapeutic agents. *Am. J. Reprod. Immunol.*, **26**, 19–22.

Bennett, C.J. (1990) Assisted reproductive techniques for the anejaculatory male. *Sem. Reprod. Endocrinol.*, **8**, 265–271.

Birkenfeld, A. and Kase, N.G. (1991) Menopause medicine: current treatment options and trends. *Comp. Ther.*, **17**, 36–45.

Birkenfeld, A. and Navot, D. (1991) Endometrial cultures and their application to new reproductive technologies. A look ahead. *In Vitro Fertil. Embryo Transfer*, **8**, 119–126.

Brackett, B.G., Seidel, G.E. and Seidel, S.M. (eds) (1981) *New Techniques in Animal Breeding*. Academic Press, NY.

Carr, J.S. and Reid, R.J. (1990) Ovulation induction with GnRH. *Sem. Reprod. Endocrinol.*, **8**, 174–185.

Cohen, J. (1991) Editorial: Efficiency and efficacy of IVF and GIFT. *Hum. Reprod.*, **6**, 613–618.

Cohen, J., Fehilly, C.B. and Edwards, R.G. (1986) Alleviating human infertility, in *Reproduction in Mammals Vol. 5* (eds C.R. Austin and R.V. Short) Cambridge University Press, pp. 148–175.

Cohen, J., Malter, H.E., Talansky, B.E. and Grifo, J. (1992) *Micromanipulation of Human Gametes and Embryos*. Raven Press, NY.

Collins, R.L. (eds) (1991) *Ovulation Induction*. Springer Verlag, NY.

Courot, M. and Volland-Nail, P. (1991) Management of reproduction in livestock: present and future. in *On the Eve of the 3rd Millennium, the European Challenge for Animal Production*. EAAP Publication No. 48 Pudoc, Wageningen, pp. 23–38.

Crosignani, P.G. and Rubin, B.L. (eds) (1983) *In Vitro Fertilization and Embryo Transfer*. Academic Press, London.

Damewood, M.D. (ed) (1990) *The Johns Hopkins Handbook of In Vitro Fertilization and Assisted Reproductive Technologies*. Little, Brown, Boston.

Davies, M.C. (1991) Pregnancy without ovaries. *Contemp. Rev. Obstet. Gynaecol.*, **3**, 119–125.

Dodson, W.C. (1990) GnRH analogues as adjunctive therapy in ovulation induction. *Sem. Reprod. Endocrinol.*, **8**, 198–207.

Drost, M. and Thatcher, W.W. (1992) Application of gonadotrophin releasing hormone as therapeutic agent in animal reproduction. *Anim. Reprod. Sci.*, **28**, 11–19.

Durrant, B.S., Oosterhuis, J.E. and Hoge, M.L. (1986) The application of artificial reproduction techniques to the propagation of selected endangered species. *Theriogenology*, **25**, 25–32.

Ebert, K.M. and Selgrath, J.P. (1991) Changes in domestic livestock through genetic engineering, in *Animal Applications of Research in Mammalian Development*. Cold Spring Harbor Laboratory Press, NY, pp. 233–266.

Edwards, R.G. (ed) (1990) Assisted Human Reproduction. *Br. Med. Bull.*, **46**, No 3.

Edwards, R.G. (1980) *Conception in the Human Female*. Academic Press, London.

First, N.L. (1991) New advances in reproductive biology of gametes and embryos. in *Animal Applications of Research in Mammalian Development*. Cold Spring Harbor Laboratory Press, NY, pp. 1–21.

First, N.L. and Parish, J.J. (1988) Sperm maturation and *in vitro* fertilization. *11th International Congress on Animal Reproduction and AI*, **5**, 161–168.

First, N.L. and Prather, R.S. (1991) Production of embryos by oocyte cytoblast–blastomere fusion in domestic animals. *J. Reprod. Fertil.*, **Suppl. 43**, 245–254.

Ginsburgh, J. and Hardiman, P. (1991) Ovulation induction with HMG. A changing scene. *Gynecol. Endocrinol.*, **5**, 57.

Gold, M. (1985) The baby makers. *Science*, **85**, 26–38.

Greve, T., Hyttel, P. and Weir, B.J. (eds) (1989) *Cell Biology of Mammalian Egg Manipulation, J. Reprod. Fertil.*, **Suppl. 38**.

Greve, T. and Madison, V. (1991) *In vitro* fertilization in cattle: a review. *Reprod. Nutr. Dev.*, **31**, 147–157.

Hafez, E.S.E. (ed) (1991) *Assisted Human Reproductive Technology. Reproductive Health Technology.* Hemisphere Publ. Co., NY.

Hamberger, L. and Wikland, M. (eds) (1992) *Assisted Reproduction. Baillière's Clin. Obstet. Gynaecol.*, 6, No 2.

Hammerstedt, R.H., Graham, J.K. and Nolan, J.P. (1990) Cryopreservation of mammalian sperm: What we ask them to survive. *J. Androl.*, 11, 73–88.

Haresign, W. (1990) Controlled breeding in sheep, in *New Developments in Sheep Production. Occasional Publication No 14.* Br. Soc. Anim. Prod., Penicuik, UK, pp. 23–37.

Hargreaves, T.B. and Soon, T.D. (eds) (1990) *The Management of Male Infertility.* PG Publishing, Singapore.

Hillier, S.G. (1990) Ovarian manipulation with pure gonadotrophins. *J. Endocrinol.*, 127, 1–4.

Holzle, C. and Wiesing, U. (1991) *In Vitro Fertilization.* Springer-Verlag, Berlin.

Howles, C.M. and Edwards, R.G. (1988) Conception *in vitro. Res. Reprod. Wall Charts.*

Isidori, A. (1990) Gonadotrophin therapy in the male, in *Clinical IVF Forum* (eds P.L. Matson and B.A. Lieberman) Manchester University Press, pp. 140–150.

Jansen, R. and Anderson, J. (1990) New routes of gametes and embryo transfer, in *Gamete Physiology* (eds R.H. Asch, J.P. Balmaceda and I. Johnstone) Serono Symposia, Norwell, pp. 301–342.

Jöchle, W. (1984) Traces of embryo transfer and artificial insemination in antiquity and the medieval age. *Theriogenology*, 21, 80–83.

Jöchle, W. and Lamond, D.S. (1980) *Control of Reproductive Functions in Domestic Animals.* Martinus Nijhoff, The Hague.

Keenan, D., Cohen, J., Suzman, M., *et al.* (1991) Stimulation cycles suppressed with GnRH analog yield accelerated embryos. *Fertil. Steril.*, 55, 792–796.

Kennaway, D.J., Dunstan, E.A. and Staples, E.A. (1987) Photoperiodic control of the end of breeding activity and fecundity of ewes. *J. Reprod. Fertil.*, Suppl. 34, 187–199.

Kraemer, D.C. (1983) Intra- and interspecific embryo transfer. *J. Exp. Zool.*, 228, 363–371.

Kraemer, D.C. (1989) Embryo collection and transfer in small ruminants. *Theriogenology*, 31, 141–148.

Lauria, A. and Gandolfi, F. (eds) (1992) *Embryonic Development and Manipulation in Animal Production.* Portland Press, London.

Leese, H. (1988) *Human Reproduction and In Vitro Fertilization.* MacMillan, London.

Lin, J.-H. and Panzer, R. (1990) To face some female animal reproductive problems with acupuncture, in *Proceedings the 5th AAAP Animal Science Congress* Vol 2. Taipei, Taiwan, pp. 305–322.

Mashiach, S., Ben-Rafael, Z., Laufer, N. and Schenker, J.G. (eds) (1990) *Advances in Assisted Reproductive Technologies.* Plenum, NY.

Matson, P.L. and Lieberman, B.A. (eds) (1990) *Clinical IVF Forum. Current Views in Assisted Reproduction.* Manchester University Press.

Meldrum, D.R. (1990) Ovulatory induction for *in vitro* fertilization. *Sem. Reprod. Endocrinol.*, 8, 213–218.

Montgomery, G.W., McNatty, K.P. and Davis, G.H. (1992) Physiology and molecular genetics of mutations that increase ovulation rate in sheep. *Endocrine Rev.*, 13, 309–328.

Niemann, H. (1991) Cryopreservation of ova and embryos from livestock: current status and research needs. *Theriogenology*, 35, 109–124.

Niemann, H. (1991) Reproductive biotechnology: Prospects and applications in the herd management of sows. *Reprod. Domestic Anim.*, **26**, 22–26.

Oehringer, S. and Alexander, N.J. (1991) Male infertility. The focus shifts to sperm manipulation. *Curr. Opin. Obstet. Gynecol.*, **3**, 182–190.

Paisley, L.A., Mickelsen, W.D. and Anderson, P.D. (1986) Mechanism and therapy for retained fetal membranes and uterine infections of the cow: a review. *Theriogenology*, **25**, 353–381.

Pedersen, R.A., McLaren, A. and First, N.L. (eds) (1991) *Animal Applications of Research in Mammalian Development.* Cold Spring Harbor Laboratory Press, US.

Petters, R.M. (1991) Embryo development *in vitro* to the blastocyst stage in cattle, pigs and sheep. *Anim. Reprod. Sci.*, **28**, 415–421.

Polge, C. (1990) Potential impact of advanced biotechnology on genetic conservation programmes, in *Genetic Conservation of Domestic Livestock* (ed L. Alderson) CAB International, Oxford, pp. 227–235.

Polge, C. (1991) Novelties in reproductive biotechnology, in *European Embryo Transfer Association: 7th Scientific Meeting Cambridge.* INRA, Nouzilly, pp. 103–116.

Prather, R.S. and Robe, J.M. (1991) Cloning by nuclear transfer and embryo splitting in laboratory and domestic animals, in *Animal Applications of Research in Mammalian Development.* Cold Spring Harbor Laboratory Press, US, pp. 205–232.

Rall, W.F. (1991) Guidelines for establishing animal genetic resource banks. Biological materials, management and facility considerations, in *Proceedings of the Wild Cattle Symposium* (eds D.L. Armstrong and T.S. Gross) Henry Doorly Zoo, Omaha, pp. 96–106.

Rall, W.F. (1992) Cryopreservation of oocytes and embryos: methods and applications. *Anim. Reprod. Sci.*, **28**, 237–245.

Reiss, H. (1991) Management of tubal infertility in the 1990s. *Br. J. Obstet. Gynaecol.*, **98**, 619–623.

Rexroad, C.E. (1989) Co-culture of domestic animal embryos. *Theriogenology*, **31**, 105–114.

Ruane, J. (1988) Review of the use of embryo transfer in the genetic improvement of dairy cattle. *Anim. Breed. Abs.*, **50**, 437–446.

Seager, S.W.J., Wildt, D.E. and Platz, C.C. (1980) Artifical breeding in captive wild animals and its possible future use, in *Current Therapy in Theriogenology* (ed D.A. Morrow) Saunders, Philadelphia, pp. 1151–1153.

Seidel, G.E. (1991) Embryo transfer: the next 100 years. *Theriogenology*, **35**, 171–180.

Seidel, G.E. (1992) Overview of cloning animals by nuclear transplantation. *Proceedings Symposium in Cloning Mammals by Nuclear Transplantation.* CSU Press, Fort Collins, pp. 1–16.

Short, R.E., Staigmiller, R.B. and Bellows, R.A. (1988) Hormonal treatments to induce ovulation. *11th International Congress on Animal Reproduction and AI*, **5**, 146–154.

Silver, L.M. (1990) New reproductive technologies in the treatment of human infertility and genetic disease. *Ther. Med.*, **11**, 103–110.

Silver, M. and Fowden, A.L. (1988) Induction of labour in domestic animals: endocrine changes and neonatal viability, in *The Endocrine Control of the Fetus* (eds W. Kunzell and A. Jensen) Springer-Verlag, Heidelberg, pp. 401–411.

Silver, M. (1990) Prenatal maturation, the timing of birth and how it may be regulated in domestic animals. *Exp. Physiol.*, **75**, 285–307.

Silverberg, K.M. and Hill, G.A. (1991) Reproductive surgery vs. assisted reproductive technologies. Selecting the correct alternative. *J. Gynecol. Surg.*, **7**, 67–82.

Sitruk-Ware, R. and Utian, W.H. (eds) (1991) *The Menopause and Hormonal Replacement Therapy*. Marcel Dekker, NY.

Smith, J.F. (1985) Immunization of ewes against steroids – a review. *Proc. NZ Soc. Anim. Prod.*, **45**, 171–177.

Stangel, J.J. (ed) (1990) *Infertility Surgery. A Multimethod Approach in Female Reproductive Surgery*. Appleton Lange, US.

Stewart, C.L. (1991) Prospects for the establishment of embryonic stem cells and genetic manipulation of domestic animals, in *Animal Applications of Research in Mammalian Development*. Cold Spring Harbor Laboratory Press, NY, pp. 267–283.

Summers, P.M. and Hearn, J.P. (1989) Embryo manipulation for the regulation of reproduction and disease, in *Reproduction and Disease in Captive and Wild Animals* (eds G.R. Smith and J.P. Hearn) *Symp. Zool. Soc. Lond. No 60*. Clarendon Press, Oxford, pp. 119–134.

Toyoda, Y. and Naito, K. (1990) IVF in domestic animals, in *Fertilization in Mammals* (eds B.D. Bavister, J. Cummins and E.R.S. Roldan) Serono Symposia, Norwell, pp. 335–347.

Trounson, A. (1992) The production of ruminant embryos *in vitro*. *Anim. Reprod. Sci.*, **28**, 125–137.

van Blerkom, J. (1991) Cryopreservation of the mammalian oocyte, in *Animal Applications of Research in Mammalian Development*. Cold Spring Harbor Laboratory Press, US, pp. 83–119.

Walters, J.R. and Hooper, P.N. (1990) Artifical insemination and its use in the conservation of genetic resources in pig breeds, in *Genetic Conservation of Domestic Livestock* (ed L. Alderson) CAB International, Oxford, pp. 221–226.

Ward, K.A. (1991) The application of transgenic techniques for the improvement of domestic animal productivity. *Curr. Opin. Biotechnol.*, **2**, 834–839.

Whitehead, M. and Godfree, V. (1991) *Hormone Replacement Therapy*. Churchill Livingstone, Edinburgh.

Wildt, D.E. (1989) Strategies for the practical application of reproductive technologies to endangered species. *Zoo Biol.*, (**Suppl**) 17–20.

Wildt, D.E. (1990) Potential applications of IVF technology for species conservation, in *Fertilization in Mammals* (eds B.D. Bavister, J. Cummins and E.R.S. Roldan) Serono Symposia, Norwell, pp. 349–364.

Wildt, D.E. (1992) Genetic resource banks for conserving wild-life species: justification examples and becoming organised on a global basis. *Anim. Reprod. Sci.*, **28**, 247–257.

Wildt, D.E., Monfort, S.L., Donoghue, A.M., *et al.* (1992) Embryogenesis in conservation biology – or how to make an endangered species embryo. *Theriogenology*, **37**, 161–184.

Wilmut, I. and Clark, A.J. (1991) Basic techniques of transgenesis. *J. Reprod. Fertil.*, **Suppl. 43**, 265–275.

Wilmut, I., Hooper, M.L. and Simons, J.P. (1991) Genetic manipulation of mammals and its application in reproductive biology. *J. Reprod. Fertil.*, **92**, 245–279.

Wilmut, I., Haley, C.S. and Woolliams, J.A. (1992) Impact of biotechnology on animal breeding. *Anim. Reprod. Sci.*, **28**, 149–162.

Wolf, D.P., Thomson, J.A., Zelinski-Wooten, M.B. and Stouffer, R.L. (1990) *In vitro* fertilization and embryo transfer in non-human primates: the technique and its applications. *Mol. Reprod. Dev.*, **27**, 261–280.

Yovich, J. and Grudzinskas, G. (1990) *Management of Infertility. A Manual of Gamete Handling Procedures*. Butterworth-Heinemann, Oxford.

Yuen, B.H. and Pride, S. (1990) Induction of ovulation with exogenous gonadotropins in anovulatory infertile women. *Sem. Reprod. Endocrinol.*, **8**, 186–197.

Chapter 7

Ada, G.L. and Griffin, P.D. (eds) (1991) *Vaccines for Fertility Regulation*. Cambridge University Press.

Alexander, N.J., Griffin, D., Spieler, J.M. and Waites, G.M.H. (eds) (1991) *Gamete Interaction: Prospects for Immunocontraception*. Wiley Liss, NY.

Althaus, F.A. and Kaeser, L. (1990) At the pill's 30th birthday, breast cancer question is unresolved. *Fam. Plan. Perspect.*, **22**, 173–176.

Anon (1990) Contraceptive efficacy of testosterone-induced azoospermia in normal men. *Lancet*, **336**, 955–959.

Avrech, O.M., Golan, A., Weinraub, Z., *et al.* (1991) Mifepristone (RU 486) alone or in combination with a PG analogue for termination of early pregnancy. A review. *Fertil. Steril.*, **56**, 385–393.

Bain, J. (1989) Male contraception. *Adv. Contraception*, **5**, 263–269.

Baulieu, E.-E. (1990) Editorial: RU 486 and the early nineties. *Endocrinology*, **127**, 2043–2046.

Baulieu, E.-E. and Segal, S.J. (eds) (1985) *The Antiprogestin Steroid RU 486 and Human Fertility Control*. Plenum Press, NY.

Blumenthal, P.D. (1991) Abortion. *Curr. Opin. Obstet. Gynecol.*, **3**, 496–500.

British Journal of Family Planning (1991) **Vol. 16 (4th Suppl)**.

Bromwich, P. and Parsons, A. (1990) *Contraception: The Facts*. Cambridge University Press.

Bucksee, K., Kumar, S. and Saraya, L. (1990) Contraceptive vaginal ring. A rising star on the contraceptive horizon. *Adv Contraception*, **6**, 177–183.

Burkman, L.J. and Kruger, T.F. (1990) Strategy of vaccine development, in *Gamete Interaction. Prospects for Immunocontraception* (eds N.J. Alexander, D. Griffin, J.M. Spieler and G.M.H. Waites) Wiley Liss, NY, pp. 501–522.

Burkman, R.T. (1991) IUDs. *Curr. Opin. Obstet. Gynecol.*, **3**, 482–491.

Chi, I.C. (1991) The Nova-T IUD. A review of the literature. *Contraception*, **44**, 341–366.

Concannon, P.W. and Meyers-Wallen, V.N. (1991) Current and proposed methods for contraception and termination of pregnancy in dogs and cats. *J. Am. Vet. Med. Assoc.*, **198**, 1214–1225.

Concannon, P.W., Morton, D.B. and Weir, B.J. (eds) (1989) *Dog and Cat Reproduction, Contraception and Artifical Insemination. J. Reprod. Fertil.*, **Suppl. 39**.

Connell, E.B. (1991) Barrier contraceptives, spermicides and periodic abstinence. *Curr. Opin. Obstet. Gynecol.*, **3**, 477–481.

Darney, P.D. (1991) Subdermal progestin implant contraception. *Curr. Opin. Obstet. Gynecol.*, **3**, 470–476.

Davis, J.E. (1990) Male sterilization. *Curr. Opin. Obstet. Gynecol.*, **2**, 535–540.

Derman, R.J. (1990) Oral contraceptives and cardiovascular risks: taking a safe course of action. *Postgrad. Med.*, **88**, 119–125.

Devlin, M.C. and Barwin, B.N. (1989) Barrier contraception. *Adv. Contraception*, **5**, 197–204.

Drife, J. (1990) Benefits and risks of oral contraceptives. *Adv. Contraception*, **6 (Suppl)**, 15–25.

Fathalla, M.E., Rosenfield, A. and Indriso, C. (eds) (1990) *The F.I.G.O. Manual of Human Reproduction, Vol. 2, Family Planning*. Parthenon, Carnforth.

Faundes, A. and Pinotti, J.A. (1990) Future methods in contraceptive research. *Curr. Opin. Obstet. Gynecol.*, **2**, 541–547.

Fraser, I.S. (1991) A review of the use of progesterone-only minipills for contraception during lactation. *Reprod. Fertil. Dev.*, **3**, 245–254.

Glasier, A. and Baird, D.T. (1990) Post-ovulatory contraception. *Baillière's Clin. Obstet. Gynaecol.*, **4**, 283–292.

Grubb, G.S. (1991) Experimental methods of contraception. *Curr. Opin. Obstet. Gynecol.*, **3**, 491–495.

Guillebaud, J. (1991) Contraception after pregnancy. *Br. J. Family Planning*, **16 (Suppl. 4)**, 16–20.

Guillebaud, J. (1991) *The Pill*, 4th Edn. Oxford University Press.

Harlap, S. (1991) Oral contraceptives and breast cancer. Cause and effect. *J. Reprod. Med.*, **36**, 374–395.

Healy, D.L. (1990) Progesterone receptor antagonists and prostaglandins in human fertility regulation: a clinical review. *Reprod. Fertil. Dev.*, **2**, 477–490.

Henshaw, S.K. (1990) Induced abortion: a world review, 1990. *Fam. Plan. Perspec.*, **22**, 76–89.

Huggins, G.R. and Cullins, V.E. (1990) Fertility after contraception or abortion. *Fertil. Steril.*, **54**, 559–573.

Katz, D.F. (1991) Human cervical mucus: research update. *Am. J. Obstet. Gynecol.*, **165**, 1984–1986.

Kessel, E., Zipper, J. and Mumford, S.D. (1988) Quinacrine non-surgical female sterilization: a reassessment of safety and efficacy, in *Modern Trends in Infertility and Conception Control 4* (eds E.E. Wallach and R.D. Kempers) Year Book Medical Publishers, Chicago, pp. 567–572.

King, R.J.B. (1991) Biology of female sex hormone action in relation to contraceptive agents and neoplasia. *Contraception*, **43**, 527–542.

Kirkpatrick, J.F., Lin, I.K.M. and Turner, J.W. (1990) Remotely delivered immunocontraception in feral horses. *Wildl. Soc. Bull.*, **18**, 326–330.

Kleinman, R.L. (1988) *Directory of Hormonal Contraceptives*. IPFF, London.

Kleinman, R.L. (1988) *Family Planning Handbook for Doctors*. IPFF, London.

Knuth, V.A. and Nieschlag, E. (1987) Endocrine approaches to male fertility control. *Baillière's Clin. Endocrinol. Metab.*, **1**, 113–131.

Landingham, M., Trussell, J. and Grummer-Strawn, L. (1991) Contraceptive and health benefits of breast feeding: A review of the recent evidence. *Int. Fam. Planning Perspect.*, **17**, 131–136.

Leiper, M.A. (1989) Preliminary evaluation of Reality, a condom for women to wear. *Adv. Contraception*, **5**, 229–235.

Limpaphayom, K. (1991) Sterilization. *Curr. Opin. Obstet. Gynecol.*, **3**, 501–510.

Mauldin, W.P. and Segal, S.J. (1989) Historical perspectives on the introduction of contraceptive technology, in *Demographic and Programmatic Consequences of Contraceptive Innovation* (eds S.J. Segal, A.O. Tsui and S.M. Rogers) Plenum Press, NY, pp. 33–52.

McGonigle, K.F. and Huggins, G.R. (1991) Oral contraceptives and breast disease. *Fertil. Steril.*, **56**, 799–819.

McPherson, K. (1991) Combined progestin and estrogen (oral) and other forms of hormonal contraceptive. *Curr. Opin. Obstet. Gynecol.*, **3**, 486–490.

Mosse, J. and Heaton, J. (1990) *The Fertility and Contraception Book*. Faber, London.

Odlind, V. (1991) Hormonal long-acting methods of contraception. *Br. J. Fam. Plan.*, **16 (Suppl. 4)**, 8–11.

Paterson, M. and Aitken, R.J. (1990) Development of vaccines targeting the zona pellucida. *Curr. Opin. Immunol.*, **2**, 743–747.

Potts, M. and Diggory, P. (1983) *Textbook of Contraceptive Practice*, 2nd Edn. Cambridge University Press.

Ray, S., Verma, P. and Kumar, A. (1991) Development of male-fertility regulatory agents. *Med. Res. Rev.*, **11**, 437–472.

Robertson, W.H. (1990) *Illustrated History of Contraception. A Concise Account of the Quest for Fertility Control.* Parthenon Publications, Lancaster.

Romieu, I., Berlin, J.A. and Colditz, G. (1990) Oral contraceptives and breast cancer. Review and meta-analysis. *Cancer*, **66**, 2253–2263.

Segal, S.J., Alvarez-Sanchez, F., Brache, V., *et al.* (1991) Norplant implants: the mechanism of contraceptive action. *Fertil. Steril.*, **56**, 273–277.

Segal, S.J. and Mauldin, W.P. (1987) Contraceptive choices, who, what, why? In *Fertility Regulation Today and Tomorrow* (eds E. Diczfalusy and M. Bygdeman) Raven Press, NY, pp. 305–317.

Simpson, J.L. and Phillips, O.P. (1990) Spermicides, hormonal contraception and congenital malformations. *Adv. Contraception*, **6**, 141–167.

Singh, K. and Ratnam, S.S. (1991) New developments in contraceptive technology. *Adv. Contraception*, **7**, 137–159.

Singh, K., Viegas, O.A. and Ratnam, S.S. (1989) Long-acting progestins. An update. *Adv. Contraception*, **5**, 241–251.

Skegg, D.C.G. (1988) Potential for bias in care-control studies of oral contraceptives and breast cancer. *Am. J. Epidemiol.*, **127**, 205–212.

Spicer, D.V., Shoupe, D. and Pike, M.C. (1991) GnRH agonists as contraceptive agents: predicted significantly reduced risk of breast cancer. *Contraception*, **44**, 289–310.

Staffa, J.A., Newschaffer, C.J., Jones, J.K. and Miller, V. (1992) Progestins and breast cancer: an epidemiologic review. *Fertil. Steril.*, **57**, 473–491.

Steel, S.J. and Guillebaud, J. (1992) Critical review of operations for female sterilization, in *Recent Advances in Obstetrics and Gynaecology*, **17** (ed J. Bonnar) Churchill Livingstone, Edinburgh, pp. 123–138.

Steinberg, W.M. (1989) Oral contraception: risks and benfits. *Adv. Contraception*, **5**, 219–228.

Stewart, G.K. (1990) Female sterilization. *Curr. Opin. Obstet. Gynecol.*, **2**, 531–534.

Stubblefield, P.G. (1990) Hormonal contraception. *Curr. Opin. Obstet. Gynecol.*, **2**, 516–523.

Sudlow, C.L.M. (1991) The contraceptive effect of breast feeding. *Br. J. Fam. Plan.*, **17**, 56–59.

Szarewski, A. (1991) *Hormonal Contraception: A Woman's Guide*. Optima, London.

Szarewski, A. and Guillebaud, J. (1991) Contraception. *Br. Med. J.*, **302**, 1224–1225.

Talwar, G.P. and Singh, D. (1988) Birth control vaccines inducing antibodies against chorionic gonadotropin, in *Contraception Research for Today and the Nineties* (ed G.P. Talwar) Springer-Verlag, NY, pp. 183–197.

Talwar, G.P., Singh, D., Pal, R. and Aruran, K. (1990) Experiences of the anti-hCG vaccine of relevance to development of other birth control devices, in *Gamete Interaction. Prospects for Immunocontraception* (eds N.J. Alexander, D. Griffin, J.M. Spieler and G.M.H. Waites) Wiley Liss, NY, pp. 579–594.

Thomas, D.B. (1991) Oral contraceptives and breast cancer: a review of the epidemiologic literature. *Contraception*, **43**, 597–642.

Thorogood, M. (1991) Cardiovascular risk of oral contraceptives. *Br. J. Fam. Plan.*, **16 (Suppl. 4)**, 30–33.

Tietz, C. and Henshaw, S L. (1986) *Induced Abortion: A World Review.* 6th Edn. Alan Guttmacher Institute, NY.

Tsui, O.A. and Herbertson, M.A. (eds) (1989) *Dynamics of Contraceptive Use. J. Biosocial. Sci.,* **Suppl. 11.**

Van Look, P.F.A. and Bygdeman, M. (1989) Antiprogestational steroids: a new dimension in human fertility regulation. *Oxf. Rev. Reprod. Biol.,* **11,** 1–60.

Van Os, W.A. (1987) IUD past, present and future. *Adv. Contraception,* **5,** 205–212.

Waites, G.M.H. (1988) *Male infertility regulation. Research in Human Reproduction.* WHO Biannual Report 1986–1987. WHO: Geneva, pp. 153–173.

Wang, M.-N. and Heap, R.B. (1992) Vaccination against pregnancy. *Oxf. Rev. Reprod. Biol.,* **14,** 101–140.

Williams, J.K. (1991) Oral contraceptives and reproductive system cancer. Benefits and risks. *J. Reprod. Med.,* **36 (Suppl),** 247–252.

Woolley, R.J. (1991) Contraception. A look forward Part 2. Mifepristone and gossypol. *J. Am. Board Fam. Pract.,* **4,** 103–113.

Wynn, V. (1991) Oral contraceptives and coronary heart disease. *J. Reprod. Med.,* **36 (Suppl),** 19–25.

Yang, M., Prasad, R.N.V. and Ratnam, S.S. (1990) Contraception: barriers and spermicides, periodic abstinence and IUDs. *Curr. Opin. Obstet. Gynecol.,* **2,** 524–530.

Chapter 8

Andrews, L.B. (1990) The evaluation of laws surrounding *in vitro* fertilization. *Sem. Reprod. Endocrinol.,* **8,** 293–303.

Austin, C.R. (1990) Dilemmas in human IVF practice, in *Fertilization in Mammals* (eds B.D. Bavister, J. Cummins, E.R.S. Roldan) Serono Symposia, Norwell, pp. 373–379.

Baird, D.T., Ledger, W.L. and Glasier, A.F. (1990) Induction of ovulation – cost-effectiveness and future prospects. *Baillière's Clin. Obstet. Gynecol.,* **4,** 639–650.

Bancroft, J. and Sartorius, N. (1990) The effects of oral contraceptives on well-being and sexuality. *Oxf. Rev. Reprod. Biol.,* **12,** 57–92.

Bartels, D.M., Priester, R., Vaiwter, D.E. and Caplan, A.L. (eds) (1990) *Beyond Baby M: Ethical Issues in New Reproductive Techniques.* Humana Press Inc, NJ.

Bromham, D.R., Dalton, M.E., Jackson, J.C. and Millican, P.J.R. (1992) *Ethics in Reproductive Medicine.* Springer-Verlag, London.

Campbell, A.V. (1990) Ethics and animal production. *Proc. NZ Soc. Anim. Prod.,* **50,** 211–213.

Cohen, J. (1991) Management of infertility: defence of medically assisted pro-creation. *Contraception Fertil. Sexual.,* **19,** 561–564.

Cohen, S. and Taub, N. (1989) *Reproductive Laws for the 1990s.* Humana Press Inc, NJ.

Djerassi, C. (1981) *The Politics of Contraception.* Freeman, Oxford.

Gillespie, D.G., Cross, H.E., Crowley, J.G. and Radloff, S.R. (1989) Financing the delivery of contraceptives: The challenge of the next twenty years, in *Demographic and Programmatic Consequences of Contraceptive Innovations* (eds S.J. Segal, A.O. Tsui and S.M. Rogers) Plenum, NY, pp. 265–295.

Grobstein, C. (1990) Ethical considerations applicable to reproductive techniques, in *Gamete Physiology* (eds R.H. Asch, J.P. Balmaceda and I. Johnston) Serono Symposia, Norwell, pp. 239–242.

McNeil, M., Varcoe, I. and Yearley, S. (1990) *The New Reproductive Technologies*. MacMillan Press, London.

Office of Technology Assessment. Congress of the United States (1988) *Infertility – medical and social choices*. US Government Printing Office, Washington.

Page, H. (1989) Economic appraisal of *in vitro* fertilization: Discussion paper. *J. R. Soc. Med.*, **82**, 99–102.

Pauncefort, Z. (1984) *Choices in Contraception*. Pan, London.

Schroten, E. (1992) Embryo production and manipulation: ethical aspects. *Anim. Reprod. Sci.*, **28**, 163–169.

Scott, R.L. (1990) Law and reproductive technology, in *Gamete Physiology* (eds R.H. Asch, J.P. Balmaceda and I. Johnston) Serono Symposia, Norwell, pp. 243–253.

Segal, S.J., Tsui, A.O. and Rogers, S.M. (eds) (1989) *Demographic and Programmatic Consequences of Contraceptive Innovations*. Plenum Press, NY.

Singer, P. and Wells, D. (1985) *Making Babies: The New Science and Ethics of Conception*. Scribners, NY.

Smith, R.L. (1976) Ecological genesis of endangered species: The philosophy of preservation. *Annu. Rev. Ecol. Systematics*, **7**, 33–55.

Templeton, A.A. and Cusine, D. (eds) (1990) *Reproductive Medicine and the Law*. Churchill Livingstone, Edinburgh.

Tendler, M.D. (1990) Religious issues in reproductive technologies, in *Gamete Physiology* (eds R.H. Asch, J.P. Balmaceda and I. Johnston) Serono Symposia, Norwell, pp. 255–259.

Wallach, E.E. (1990) Ethical aspects of IVF and related assisted reproductive technologies, in *The Johns Hopkins Handbook of In Vitro Fertilization and Assisted Reproductive Technologies* (ed M.D. Damewood) Little, Brown, Boston, pp. 151–162.

Warnock, M. (1985) *A Question of Life*. Blackwell Scientific Publications, Oxford.

Winston, R.M.L. (1991) Resources for infertility treatment. *Baillière's Clin. Obstet. Gynaecol.*, **5**, 551–574.

Index